D1004738

NANO COMES TO LIFE

NANO COMES TO LIFE

How Nanotechnology Is Transforming Medicine and the Future of Biology

Sonia Contera

PRINCETON UNIVERSITY PRESS
PRINCETON AND OXFORD

Published by Princeton University Press
41 William Street, Princeton, New Jersey 08540
6 Oxford Street, Woodstock, Oxfordshire OX20 1TR

press.princeton.edu

Library of Congress Cataloging-in-Publication Data

Names: Contera, Sonia, author.
Title: Nano comes to life : how nanotechnology is transforming medicine
and the future of biology / Sonia Contera.
Description: Princeton : Princeton University Press, [2019] |
Includes bibliographical references and index.
Identifiers: LCCN 2019023184 (print) | LCCN 2019023185 (ebook) |
ISBN 9780691168807 (hardback ; alk. paper) | ISBN 9780691189284 (ebook)
Subjects: MESH: Nanotechnology
Classification: LCC QH324.25 (print) | LCC QH324.25 (ebook) |
NLM QT 36.5 | DDC 570.285–dc23
LC record available at https://lccn.loc.gov/2019023184
LC ebook record available at https://lccn.loc.gov/2019023185

British Library Cataloging-in-Publication Data is available

Editorial: Ingrid Gnerlich and Arthur Werneck
Production Editorial: Kathleen Cioffi
Text Design: Carmina Alvarez
Jacket Design: Jason Alejandro
Production: Jacqueline Poirier
Publicity: Sara Henning-Stout and Katie Lewis
Copyediting: Annie Gottlieb

Jacket image: Nano-vaccine cancer treatment, SEM © National Cancer Institute

This book has been composed in Minion Pro

Printed on acid-free paper. ∞

Printed in the United States of America

10 9 8 7 6 5 4 3 2 1

FOR ARTURO AND ISADORA

Contents

Preface and Acknowledgments

The progressive convergence of sciences in the twenty-first century, and in particular, the merging of disciplines at the interface of physics, nanotechnology, biology and medicine, has composed the landscape of my own scientific career across sciences, continents, and cultures. After a study and work journey that led me from physics to nanotechnology to biology and back to physics, through Spain, China, Czechia, Japan, Denmark and the UK, in 2007 I became the co-director of the Institute of Nanoscience for Medicine, a research program at the Oxford Martin School of the University of Oxford. The school was created with an endowment from James and Lillian Martin to become a hub where all the relevant academic disciplines would convene to investigate and debate the challenges and opportunities of the twenty-first century. Encouraged by the Oxford Martin School's founding mandate to communicate with the public, I started to deliver public lectures about nanotechnology and the future of medicine and biology that were strongly rooted in my physicist's way of looking at the world. Despite the quickening pace of scientific convergence, the scientific community has been slower to reflect on how the merging of disciplines is transforming the ways we work and think about nature, so my lectures were also attempts to satisfy my own needs as a practitioner of science. Speaking about these issues in public to scientific and nonscientific audiences has become an important part of my academic activity, and has led me to reflect more on the implications, history, and context of my research. I now deliver these lectures in many countries and to a wide variety of audiences. This has allowed me to connect with many communities and to become aware of the public's great curiosity about these converging technologies that so define our present and will most likely shape our future.

So when I was approached by my editor, Ingrid Gnerlich, to write this book, I decided to do so despite having a heavy academic and research load and two small children. People of all backgrounds seem to enjoy the scientific stories I tell. We are living in exciting times; breakthroughs in our understanding of the physical and biological reality around and within us are speeding up exponentially. The convergence of the sciences is bringing a revolution not only in technology, but also in our physical, cultural, and philosophical relationship to the material world. It is a time to think and talk about the fast-changing present, and to collectively imagine positive futures our new technologies make possible. I hope that this book will contribute to the conversation in a meaningful way.

I am grateful for the support and patience of my family, and for the kind encouragement of my editor; I am grateful, also, to the friends and colleagues that have read and commented on the first versions of the manuscript: Charles Olsen, Rosario Ruibal Villaseñor, Alberto Merchante, Ibon Santiago, and Lina Gálvez. I have also benefited from the generosity of Iwan Schaap, and of teamLab, who gave me beautiful images and inspired some of the ideas in the book. Many conversations have been important in shaping my thinking here, especially those with the physicist Jacob Seifert, my PhD supervisor Hiroshi Iwasaki, the film director Alison Rose, and the historian of mathematics Agathe Keller.

Abbreviations

DNA:	deoxyribonucleic acid
AI:	artificial intelligence
PET:	positron emission tomography
Length units:	m, meter; mm, millimeter; cm, centimeter; nm, nanometer
RNA:	ribonucleic acid
A, C, G, T:	adenine, cytosine, guanine, and thymine, the DNA bases
STM:	scanning tunneling microscopy/microscope
AFM:	atomic force microscopy/microscope
C:	degree Celsius (temperature)
ATP:	adenosine triphosphate
3-D:	three-dimensional
NMR:	nuclear magnetic resonance
MRSA:	methicillin-resistant *Staphylococcus aureus*
FDA:	U.S. Food and Drug Administration

NANO
COMES
TO LIFE

INTRODUCTION

SCIENCES CONVERGE IN BIOLOGY TO TRANSFORM HEALTH

Biology is the most intensely investigated subject of modern science. Beyond perpetual human preoccupations with health, mortality, and finding our place and identity in the universe, the power hidden in biology's complexity is causing almost all the branches of science and technology to gravitate toward the study of life. Biology ceases to be the sovereign territory of biologists, biochemists, and medical scientists; in the twenty-first century, physical, mathematical, and engineering sciences converge with the more traditional biological disciplines to seek a deeper understanding of life in all its multifaceted, dynamic structures and functions. In our turbulent and disoriented times, the inner workings of biology and its profound insight into the meaning of life have become the focus of human creativity, spawning technological and cultural innovations that may contribute either to our survival or to our extinction.

The sciences' appetite for biology seeks satisfaction on all its spatial scales—from nanometer-size molecules to cells tens of micrometers large to meter-scale eukaryotes[1]—and in all its manifestations, from the mind-boggling diversity of shape and action found in its molecular inventory to the forces and processes that drive the precise assembly of an intricate protein, lipid membrane, or coil of DNA. Science seeks knowledge about individual molecules, cells, tissues, organisms, and ecosystems; this includes the study of how biological structures give rise to the individual and collective "intelligences"[2] that enable living creatures to persist on Earth.

Apart from the pure search for knowledge, economic gain and social influence are the workaday drivers of science (and even more so of research funding); thus one can observe that the motivation of the current scientific desire for all things biological is often technological. The potential technological payoffs of biology are as diverse as the new disciplines evolving out of the knowledge extracted from it. For example, computer scientists are keen to learn the fine details of the human brain's organization so that they can mirror the layered connectivity between its neurons in the structure of their algorithms; they hope this will lead to much-improved artificial intelligence (AI) as well as to better understanding of our own thinking ability. Materials scientists and roboticists look to the assembly of biological structures for inspiration in the design of novel bioinspired materials and robots. In physics departments, scientists study the plant proteins responsible for photosynthesis, prospecting for biological recipes that can be adopted in the quantum computers of the future.

However vigorous and dedicated the biological research activity of these new players, medicine still takes center stage as the main intellectual, social, and economic engine of biological research. Medicine helps to attract the money, but more fundamentally, plays the role of integrator of knowledge. The sciences and technologies drawn to biology arrive by different paths and aim at disparate goals, but medicine dispels the cultural barriers among disciplines, facilitating their fusion in the pursuit of better strategies for uncovering the ultimate causes of disease and better interventions to preserve and restore health.

Understanding disease and curing it is such a complex challenge that it requires "all hands on deck"—all the technical and scientific knowledge available. Cutting-edge medical research already combines the latest advances in AI, materials science, and robotics, and will undoubtedly use quantum computers as they become available. As anyone who has been in a modern hospital can attest, most human technologies end up being adapted for use in the clinic in one way or another: from the humble thermometer to the physics of positrons in PET scans for imaging tumors, and from mobile phone apps to control fertility to gene editing to eradicate diseases. The hospital is the most nourishing culture medium for scientific and technical knowledge to combine and grow in.

The diversity, intensity, and speed of advance of current research unequivocally indicate that we are living in prerevolutionary times in both biology and medicine. Confident answers to the long-standing questions that have enthralled humans, such as the origin and diversity of life and the source of our intelligence and consciousness, are perhaps still far from being found. However, the accelerating and ever-more-potent interdisciplinary mergers make us feel that we are now at an inflection point, and will soon slide irrevocably toward the advent of the technologies that will transform our understanding and control of our biology. In extraordinarily novel and efficient ways, these will give us the powers to heal ourselves and to prolong and transform our lives.

NANOTECHNOLOGY IN BIOLOGY AND MEDICINE

A necessary step toward this brink of breakthrough was, and continues to be, the development of nanotechnology—the capacity to visualize, interact with, manipulate, and create matter at the nanometer scale. This is primarily because the main molecular players in biology, and the main drug and treatment targets in medicine—proteins and DNA—are nano-size. Nanotechnology is the technological interface with the nanoscale. It directly links the macroscopic world of our perceptions with the nanoscopic world of individual biomolecules. To arrive in medical heaven—the power to restore perfect health—we would need to know how molecules work in a specific environment, why and how they malfunction in a disease, and most importantly, how to reach them, to target them, and to deactivate or activate them. In this "spatial" sense, medicine parallels nanotechnology: to cure, we need to traverse the spectrum of scale from the macroscopic size of the doctor to the nanometer scale of biomolecules, navigating the very intricate "multiscale" landscape of organs, tissues, and cells in between. Since the early days of nanotechnology, one of its main missions has been to create tools that are able to interact with key biological molecules one at a time, directly in their complex medium, and in this way to bring closer to realization the targeting of individual molecules in the medical context. We are still working on it, and this book is in part an effort to show how far we have come.

As well as introducing nano-tools that enable new biological and medical research, nanotechnology has made a more fundamental contribution: attracting physical scientists to biology. In the last decades of the twentieth century, artificial nanomaterials and the tools of nanotechnology—microscopes and nano-manipulation devices—came into existence. Using them, a significant number of physical scientists interested in matter at the nanometer scale sought to know how and why biology first constructed itself using nano-size building blocks in the medium of (salty) water. Fascinated by the coupling of physics and chemistry that gives rise to biological function, they focused on using nanotechnology's methods to learn the workings of proteins, DNA, and other important nano-size biomolecules. In the process, they turned themselves into *biological physicists*, seeking answers to deep scientific questions such as: What was it about the properties of the nanoscale that made it special for the emergence of life? Others, more practical, saw opportunities to design nanomaterials that could be used to address disease in a more precise and rational manner, improving on current pharmacological treatments; they became *nanomedicine scientists.*

This cross-disciplinary activity led to the development of tools specifically built for studying biological processes and their nanoactors in physiological conditions (warm, salty water). As pioneering nano-bioscientists enlarged their knowledge of biology, they eroded the boundaries between materials sciences, physics, chemistry, and biology, emerging as a new generation of researchers who naturally worked across disciplines and no longer recognized intellectual or cultural barriers to interaction with any other scientific field.

THE EMERGENCE OF QUANTITATIVE BIOLOGY: THE NEW PHYSICS OF LIFE

The arrival of nanotechnology in the life sciences has contributed to a rising wave of physical scientists entering biology, bringing fresh eyes to old problems. The experiments of these scientists differ from most biological and biochemical research in that they are driven by mechanistic hypotheses: that is, they pursue quantitative data that help to explain the actual functioning mechanism of the process under study. The usual question of a biological scientist is, "Who [which molecule] does that?" For a physicist it is, "How and why does it do that, and can I model it with mathematics?" When you look at biological systems through the eyes of a physicist, you are looking for the key parameters that explain how the biological system works: Is it size, temperature, energy, speed, structure, stiffness, charge, chemical activity?

Crucially, the ultimate goal of physicists is to create mathematical models of biological processes that can be used to describe those mechanisms. If the mathematical model reproduces and even predicts the biology of the process, then we start to know the actual fundamental quantities and forces that drive it. The strength of this "quantitative approach" to biology is that it unleashes a formidable power: accurate mathematical models can be used to predict the behavior of specific biological processes in the computer, or in modern scientific jargon, *in silico*, without experiments. This means that, if successful, mathematical models can be used to progressively abandon the trial-and-error methods of the traditional biological, medical, and pharmacological sciences. These are painfully slow and costly, and, as the development of new drugs often shows, inefficient. The computer modeling approach is already in use in modern civil engineering, aeronautics, and architecture, where computer simulations combined with quantitative knowledge of the mechanical properties (e.g., elasticity, viscosity, strength, rigidity) of materials used in construction are routinely employed by engineers to test the feasibility of designs *in silico* before any actual building work is done.

Without the invention of techniques able to quantitatively monitor biology in all its dynamic, hierarchically structured complexity—from the nanometer scale of proteins and DNA to cells to tissues in living bodies—adopting this quantitative approach in medicine was totally impossible. These techniques not only need to visualize structures and their movements at all the different scales, but need to be able to extract the key physical or chemical parameters (stiffness, charge, temperature, etc.) that allow the development of correct mathematical models to make computer modeling viable.

Once experimental information at the nanometer scale of single molecules becomes available, it can be used to construct models that describe the functioning of, for example, proteins or DNA in their natural environment and in disease. The capacity to model individual molecules will be progressively integrated with the emergence of techniques able to collect vast amounts of quantitative data about those molecules in complex biological environments and in real time. Furthermore, AI algorithms (such as those of machine learning) will be used more and more to aid in the analysis of biological "big data."[3] The integration of biological physics with biological big data and AI models will lead to increasingly accurate and "smart" models of life. However, twentieth-century physics teaches us that in very complex and interconnected systems, knowing the workings of the building blocks is not enough to predict the behavior of the whole: at larger scales, biology exhibits behaviors that the smaller constituents do not exhibit, or that cannot be explained from the relationships between their molecular building blocks. This is because complexly organized matter presents collective phenomena arising from cooperative interactions between the building blocks—or, as we say in physics, these properties *emerge*. Some examples of emergent behavior are cellular movements, mechanical vibrations in the brain, electrical signaling across the membranes of cells, and changes in shape or stiffness, none of which can be predicted from just knowing the molecules that constitute a particular structure. This means that in practical terms, as we zoom out from the nanoscale to the microscale, nanoscale models have to be "coarse-grained" to be integrated and consistent with models that correctly describe the cellular behaviors emergent from nanometer-scale activity.

Similarly, the cellular level then needs to be integrated into models of the tissue and organ levels. An example would be a mathematical model of a tumor that is able to relate its shape, size, and growth pattern to the properties of individual tumor cells and their molecular environment; at the next level down in size, the model should incorporate how cellular properties are connected to their molecular and genetic activity. This model could in principle be used to design a multimodal treatment regimen that targets individual molecules both directly and indirectly. Combining nano-precise drug delivery with a physical treatment such as applying electrical or mechanical signals to the tumor would single out specific molecules and also affect them through the physical and chemical phenomena that link the different spatial and temporal scales of the tumor. In other words, it would allow simultaneous targeting of the molecular, the cellular, and the tissue-level biology of the tumor. The undertaking is formidable, but the tools that would make it possible are slowly being developed and coming together.

We can draw some parallels with the past. At the beginning of the twentieth century, the arrival of tools to study atoms conduced to the development of the field of quantum mechanics.[4] This, in turn, led to the very creative mathematical models underpinning

NANOSCALE (1–100 nm) Biomolecules	MICROSCALE (0.5–100 microns) Cells	MACROSCALE (mm–cm–m) Tissues (muscle, heart, skin...)
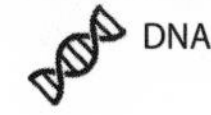 DNA	  *E coli* bacterium ~ 1 micron	
Proteins typically 3–100 nm	 Animal cell 10–100 micron (or more)	

QUANTITATIVE MEASUREMENT of movement, shape, mechanics, charge, molecules involved, what genes are active ...

Techniques (a few examples): AFM, optical tweezers, microscopies	Microscopies, genomics, proteomics, AFM	Microscopy, histology, rheology, ultrasound

MATHEMATICAL MODELS:

Models of biomolecular function	Models of cellular behavior	Models of tissue

MULTISCALE SMART MODEL INTEGRATES IT ALL!
Combines physics, modeling, simulation, and AI

Fig 0.1. The new physics of life tries to build mechanistic models of biology at each of the relevant scales, and then integrate the models into larger “multiscale” models that include all the relevant scales. (nm = nanometer)

solid-state physics, which successfully explained how the macroscopic properties of crystalline solids[5] emerged from the order and nature of their atoms. This ultimately laid the theoretical foundation for the modern electronics present in our mobile phones and other electronic devices.

While biology is immensely more complex than crystalline solids, current trends of research in all the sciences converging on biology indisputably indicate that this colossally arduous task is already under way. We are moving, still slowly, but at an inexorable pace, toward the quantitative, mathematical description of biological phenomena—in other words, *the physics of life*.

In this new landscape, the reductionist vision of the previous generation, which strove to present organisms as mere biochemical computers executing a program, an algorithm encoded in genes, has been called into serious question. Confronting the often skeptical eyes of more-traditionally trained biologists, nano- and physics- and mathematics-savvy scientists are slowly deploying their plans to quantitatively interpret the interwoven genetic, chemical, and physical mechanisms underlying life and health, and to mathematically predict the biology underlying disease and trauma. Significantly for medicine, they seek to implement their rational health-restoring strategies one patient at a time. Their final goal is to design—using mathematics and computer models—treatments for specific problems in particular patients, rather than to discover, by endless rounds of trial and error, prescriptions that work for an acceptable majority of patients, as we do now.

THE TRANSFORMATION OF BIOLOGY AND MEDICINE

In this book I seek to make sense of the reality that I am living and witnessing as a scientist working across disciplines. I am uniquely placed to tell the story of how the combined efforts of physical and mathematical scientists, facilitated by the rise of nanotechnology and of powerful quantitative experimental techniques, are transfiguring biology and slowly building up the capacity to identify and take on core challenges of modern medicine. Doing medicine means dealing with the intricate, dynamic, circuitous, hierarchical assembly of a myriad of nano-size building blocks that constitutes a living organism. To cure, we need to reach specific cells, proteins, and DNA with optimized concentrations of precisely designed therapeutic agents; to heal and regenerate, we need to understand and reproduce the nanoscale environment of healthy cells in tissues and organs. These are the subjects where nanotechnology, and the sciences and technologies enabled by it, are currently changing the game. They are the core of this book.

Chapter 1 describes how the deployment of nanotechnology and quantitative methods in biology is expanding the realm of science to embrace life in all its multiscale, entangled intricacy and multiplicity. I attempt to gather some of the most significant achievements to sketch a map of the changing landscape of biological research. This new landscape emerges, in part, from the realization that current medical bottlenecks will remain blocked if we linger in the previous generation's optimistic hope that life could be reduced to genes and molecules. As we dive into the complexity of biology, we are conscious that the ultimate goal of explaining life in all its detail is currently unattainable, but we rely on quantitative methods and new tools to provide solutions to specific problems that pave the way. I illustrate this picture with examples of how quantitative biology methods are being used to query the functions, structures, and interactions of key biological molecules, and to understand how these molecules' concerted actions give rise to subcellular structures that integrate their nanoscale activity into the emergent workings of whole cells and organisms.

Complementary to this approach, and in response to the doubt-inducing complexity of biology ("How do we really know that our models are correct?"), others have adopted a more pragmatic strategy, based on the engineering ethos of nanotechnologists and quantitative biologists. This strategy stems from the visionary physicist Richard Feynman's "What I cannot create, I do not understand"—in other words, learn by making. We know, for example, that the quantum theory of solid-state materials is accurate because our mobile phones work as we want them to. If the devices we design don't work, this unequivocally indicates that something is wrong or missing in the science we seek to apply. Application is the ultimate test of whether a scientific theory works.

The application of this principle at the interface of nanotechnology and biological sciences is transforming the way human-made materials are constructed and the uses they can have. Chapters 2 through 4 review how nanotechnology and the sciences and technologies converging on the nanometer scale are being used to create bioinspired and biomimetic nanostructures and nanomachines, and to integrate these nanostructures into strategies aimed at solving specific medical problems and clearing roadblocks.

In chapter 2, I summarize the development of disciplines that seek to build artificial nanostructures using both biological molecules and the organizational principles of biology. This activity began with the emergence of the DNA nanotechnology that is maturing as a distinct field of research. DNA nanotechnology pursues the design and construction of any arbitrary shape using artificial DNA building blocks. More importantly, DNA nanotech seeks to give its structures nanoscale functionality for complex tasks, such as serving as an active template for the synthesis of molecules, or even working as a programmable molecular computer or a DNA robot able to deliver drug payloads to tumors. In parallel, the field of protein nanotechnology attempts to achieve the same goals using natural or artificial proteins; this is a tougher task than making nanostructures with DNA, but can lead to many more applications. One of the most striking examples of the integration of sciences and the power of quantitative approaches in biology has been the arrival of "designer proteins." Recently scientists have been able to hack the molecular machinery of the cell to create "post-evolutionary" proteins that do not exist in nature, but have been previously designed in powerful computer programs. These technologies have made possible the realization of one of the dreams of the nanotechnology pioneers: the deployment of molecular assemblers able to construct any shape with atomic precision following a rational design, a plan previously rehearsed in a computer. A fundamental drive of biomolecular nanotechnology is to create powerful tools for nanomedical applications, ranging from molecular DNA assemblers of medicinal drugs to improved vaccines, powerful antiviral and antibacterial nanomedicines, and targeted drug delivery systems.

Chapter 3 gives an account of a key aspect of this multidisciplinary arsenal of nanomedicine: how nanotechnology is being used to improve the efficiency of current cancer chemotherapies by developing drug-delivery strategies that specifically target tumors and cancer cells. Although drug delivery via nanostructures was one of the initial goals of nanomedicine that attracted the most support, the effort put into it has not led to the breakthroughs that were hoped for. This is partly due to inertia: the application of existing trial-and-error methods to complex biological processes that are still poorly understood; the insufficient coordination of existing research; and the lack of truly quantitative approaches. And it is partly due to impatience: seeking nano-powered magic bullets and lucky shortcuts to cure disease that overlook the complexity of the biology involved has not proven fruitful. Fortunately, the lessons are being learned and new, improved initiatives are already gathering pace. In this chapter, I also discuss how nanotechnology is joining the multidisciplinary quest for new ways to combat antimicrobial resistance, coupling traditional pharmacology research with novel ways to interact with bacteria that include their physics, not only their chemistry. I also give a brief overview of how nanotechnology and its offshoots are becoming ever better at creating nanodevices that sense the chemicals in the body, thereby getting closer to the goal of responding to chemical imbalances in real time by releasing drugs when and where they are needed.

Perhaps one of the most fascinating contributions that nanotechnology can make to health and medicine is to team up with the biological research currently being done on immunotherapies (a type of cancer treatment that boosts the body's immune system to fight cancer). These combined efforts have the potential to accelerate the science of controlling and improving our immune system's innate capacity to detect and fight disease from within. In this chapter, I anticipate how the convergence of sciences is likely to lead to plans to create the "super-enhanced human immune system" of the future.

In chapter 4, I attempt to compile the science of one of the most potentially transformative scientific fields: tissue engineering. Tissue engineering is emerging not only as the field that may enable the repair and even replacement of damaged or diseased organs, but also as an arena where fundamental progress will be made in the basic science underpinning biology and medicine, with the goal of being able to monitor health and disease with molecular precision in real time. Studying all the relevant molecules in a large living organism is a daunting task; however, tissue engineering allows the construction of artificial biological tissues and organs, in which interactions between the scales can be tested in controlled environments. "Learning by creating" toy models of body parts and even trying to connect them in the lab will be useful for building mechanistic models that increasingly approach the complexity of real organisms. This activity will be leavened by mathematical modeling and simulation, and will likely incorporate AI (machine learning) algorithms.

Measurement and monitoring of the key parameters that can be used to create models of tissues are facilitated by the development of biosensors for constant surveillance of artificial tissues and new AI algorithms to integrate the data with the physics of tissues. Eventually, this will lead to the development of technology that may be used in vivo once it has been well established in tissue-engineering experiments. Creating biosensing technology and mechanistic models of tissues that link the molecular with micro- and macroscopic biology will arguably be the most important contributions to medicine and biology of tissue engineering. Tissue-engineering models are also very useful for understanding and modeling targeted drug delivery, and it is expected that tissue-engineered models of human tissues and organs will eventually replace animals in drug testing.

This book seeks both to describe the new science emerging from the convergence of disciplines on biology and human health and to reflect on how and why the sciences are converging. Each chapter, therefore, has a very short historical introduction outlining the path that led to the current situation. I hope this helps my purpose: to invite the reader to look back to where the science came from, in order to envisage the routes that can take us from here into the future.

TRANSMATERIAL FUTURES

Much of the science that I have briefly outlined leads to an inexorable dimming of the distinction between biological and material sciences: a new *transmaterial* science is in its embryonic state. With increasing control of matter at the nanometer scale and better knowledge of the building tricks and machinery of biology, artificial materials inspired by biology will be used to create new scaffolds for regenerating tissues and organs, or to improve the responses of the immune system. In parallel, hybrid bio-inorganic devices that mimic biology will be used in new computers and electronic devices. As biology becomes quantitative, and we gain the power of mathematics and physics to use the rules that govern it to design new applications, we release a colossal capacity for innovation, not only in medicine but in most technologies currently created by humans, from energy to electronics and from computing to materials science. By increasingly refining our ability to learn biology using the methods of physics, we will in fact be distilling the recipes of the universe to fabricate and assemble matter from the nanometer scale up, and will acquire the ultimate power to revolutionize human technology and medicine.

Forecasting the consequences of the convergence of the sciences beyond medicine (the so-called "fourth industrial revolution") is outside the scope of this book. I have, however, included an epilogue offering a scientist's perspective on how to navigate the promise and peril of a future in which we have snowballing power over all sorts of matter—biological and otherwise. Furthermore, I briefly explore the consequences for human identity (from my own scientist perspective) of the merger between material and biological sciences. As I read some of the predictive narratives on the fourth industrial revolution that have become international best sellers in recent years, I cannot avoid thinking that these books (more or less) unintentionally invite an additional danger, at least as powerful as technology's effects on society: they risk unleashing the *fear* of technology, and so undermining the power of science to create a fairer society. Much of the forecasting is based on a suboptimal knowledge of the current state of the sciences, and, more importantly, a lack of knowledge of scientists themselves—their increasing sense of vocation and commitment to engage with society, to form democratic alliances that allow positive and practical transformations for the benefit of us all.

In the twenty-first century, many scientists are passionately searching for ways to create platforms and frameworks of collaboration with the public, the regulators, and the industrial developers of technologies to imagine better, more-diverse and equable futures. Much of the writing about technology in the twenty-first century forgets that scientists, more than anyone, understand the power of the knowledge that they create, and that they increasingly strive to modulate the social and economic forces that shape its development and exploitation. Scientists are a fundamental piece in the machinery that links technology with fairness in society. While it is true that the pursuit of pure knowledge motivates many of us, and that some are motivated to build successful careers that will bring them prizes, status, and money, the reality is that most scientists endure painfully long hours in the lab or at the computer in pursuit of a deep and genuine passion to improve life for all.[6] This endeavor is actually one of the main reasons why technologies converge in medicine: contributing to better health often seems the most direct pathway for scientists to improve universal well-being, or so we hope.

This book is my attempt to convey the excitement of the new worlds that the sciences at this interface of biology, physics, and medicine are uncovering, and to share and think through with the reader the opportunities now emerging from our laboratories to use technology to collectively create a fairer future of human betterment. As I introduce in chapter 1, and reflect on further in the epilogue, the incorporation of biology (including intelligence) into the realm of physics facilitates a profound and potentially groundbreaking cultural shift, because it places the study of life within the widest possible context: the study of the rules that govern the cosmos. I want to reveal this new context for studying life and the potential for human advancement that it enables. The most powerful message of this book is that in the twenty-first century life can no longer be considered just the biochemical product of an algorithm written in genes (that can potentially be modified at someone's convenience), but a complex and magnificent realization of the laws that created the universe itself. This means that as physics, engineering, computer science, and materials science merge with biology, they are actually helping to reconnect science and technology with the deep questions that humans have asked themselves from the beginning of civilization: What is life? What does it mean to be human when we can manipulate and even exploit our own biology? We have reached a point in history where these questions naturally arise from the practice of science, and this necessarily changes the sciences' relationship with societies and cultures.

We are entering a historic period of scientific convergence, feeling an urge to turn our heads to the past even as we walk toward the future, seeking to find in the origin of the ideas that brought us here the inspiration that will allow us to move forward. This book attempts to call attention to the potential for a new intellectual framework to emerge at the convergence of the sciences, one that scientists, engineers, artists, and thinkers should tap to create narratives and visions of the future that midwife our coming of age as a technological species. This might be the most important role of the physics of life that emerges from our labs: to contribute to the collective construction of a path to the preservation of (human) life on Earth.

1

EMBRACING BIOLOGY'S COMPLEXITY, AT LAST

The microscopy technologies developed from the seventeenth century onward allowed humans to discover the basic structures of life: first the biological cell, and then, with increasing resolution, the contents of its interior, down to the unimaginable molecular complexity that characterizes life at the microscopic level.[1] By the mid-1900s, scientific discoveries led to the means to resolve the details of the structures of biomolecules with atomic precision. Simultaneously, progress in genetics and biochemistry brought increasing understanding of the building blocks of life, proteins—their structures and chemical activity, their relation to DNA and to the genetic information it contained. This gave scientists the confidence to develop an interpretation of biology that attempted an intellectual end run around its daunting convolutedness: living organisms could be explained as biochemical computers executing a molecular program, an algorithm encoded in genes—or so they hoped.

In this chapter, I will explore how science in the twenty-first century is reassessing the adequacy of this simplistic plan to reduce biology to molecules and genes. The roadblocks encountered in the biological and medical sciences, coinciding with the arrival of powerful tools that can image and interact directly with biological matter at the nanoscale, have made it both imperative and possible to question the views of the previous generation. Knowledge of genes alone is now considered insufficient to explain life or to solve the challenges of medicine. The combined efforts of quantitative experimental techniques and mathematical descriptions of biology are pushing scientists to dare to bravely embrace life in all its complexity as a "symphonic interplay between genes, cells, organs, body, and environment."[2]

Fundamental new ways of looking at life beyond individual molecules include the study of the physical mechanisms that underpin molecules' function and their assembly into "living shapes" capable of displaying very diverse behaviors at different temporal and spatial scales. These new views seek to bridge the scientific and intellectual gulfs between the nanoscale (molecules), the micron scale (cells), the millimeter and centimeter scales of tissues and organs, the meter size of large organisms, and so on, up to whole-planet-size ecosystems.

Physics has learned to accept that when systems become very complex, they often present characteristics that cannot be explained by interactions of their building blocks, because their structures and properties at larger scales *emerge* from collective behaviors at smaller scales. Biology is the clearest example: how else could the precise assembly of nanoscale building blocks eventually give rise to sentient, social beings?

Identifying the properties that are key to understanding biological complexity and its emergent behaviors has become a fundamental objective of quantitative biology. One such property is the mechanical nature of living matter. Mechanics is the branch of physics that deals with motion and the forces and mechanical properties of materials (e.g., elasticity and viscosity) that underlie the movements of simple and complex structures. The capacity to sense, respond to, and exert mechanical forces and signals is a particularly important characteristic of life at every scale that had been almost totally overlooked in the previous century. Mechanics is currently entering the biological limelight, and is doing so in concert with another physical property: biological electricity.

In this chapter, I summarize the intellectual transformation of biological research in the last decades by new techniques and concepts originating in nanotechnology, physics, and mathematical sciences. Currently a much richer, and perhaps humbler, picture is emerging as biology gradually incorporates other sciences, which merge in the search both for new knowledge and for new technologies and medical applications. As biology becomes part of the territory of physics, its new quantitative knowledge will be applied as "engineering" in medicine, materials sciences, and even computing. Toward the end of the chapter, I argue that the new "physics of life" brings us closer to a transformation of what we understand by "matter" in general, and consequently to a technological leap in what we can do with it. More importantly, placing biology within the realm of physics changes scientific culture. Physics urges us to consider life as a whole emergent from the greater whole—emanating from the same rules that govern the entire cosmos.

HIERARCHICAL UNIVERSE, HIERARCHICAL LIFE

The opening scene of the legendary short film *Powers of Ten*, created by Ray and Charles Eames[3] in the 1960s, shows a 1-square-meter (1 meter $= 10^0$ m) overhead image of a man and a woman enjoying their picnic on a blanket, near Chicago. After ten seconds, the viewpoint starts to move out and away from the blanket, at a speed of a factor of 10 in size every ten seconds: from 1 (10^0m) square meter of the picnic blanket to 10 (10^1) square meters of Burnham Park; then it continues to move at this speed to the whole city of Chicago (10^4), followed by the Earth, the Solar System, our galaxy . . . all the way to 10^{24} square meters, a snapshot of the span of the observable universe. The film takes the viewer on a cosmic journey through magnitude to discover the effect of adding another zero to the field of vision. The effect on the viewer is wonder at the hierarchical structure of the physical reality of our universe across its temporal and spatial scales, which invites reflection on the fabric of reality. (See plate 1.)

In the second part of the film, the camera zooms back in on the man on the picnic blanket to start descending into negative powers of ten, starting from 10^{-1} m ($=10$ cm): the back of the man's hand. The viewer is then taken through the skin of his hand to focus on the most complex of all hierarchical constructions: those in the biology of his cells, their subcellular structures, and eventually, his DNA. But we don't stop there. The film continues to zoom in: to one of the carbon atoms in the DNA, to the nucleus of that atom, and finally, to the quarks vibrating inside its protons at 10^{-16} m.

Navigating through the shapes underpinning life, the film highlights the discovery that biology is also arranged in a hierarchy of scales, resembling—in the layered nature of its structures—the rest of the universe it emerged from. Biology differs radically in one key feature, however: life's complexity of structure, variation, and behavior far surpasses that of any other assemblage so far known in the universe.

The appearance of nanometer-size molecules in the water of the young planet Earth around 4 billion years ago[4] generated matter that was destined for mind-blowing complexity and activity, unlike the simpler order emerging from inert quarks, atoms, or stars in space. Somehow, some of those early nanoscale molecules vibrating in salty water (or salty ice!)—the primordial RNAs and proteins—became able to use the energy they dissipated into their environment to become more complex and more ordered, to interact and self-assemble into increasingly sophisticated shapes and structures that eventually became capable of replication.[5]

With the appearance of early replicator molecules that were able to grow and divide, there arrived the key property that sets biological matter apart from any other matter known in the universe: evolution. Evolution is the universe's mechanism for fine-tuning the physics and chemistry of nanoscale molecular building blocks to create increasingly refined living structures. The physics of the universe ensures that biological complexity naturally increases in molecules that dissipate energy into their environment, while evolution *computes* the form and activity of the organisms that are able to succeed in that environment. In this context, by "computation" I mean that life embeds in itself the capacity to "compute" organisms that are able to survive in a complexly changing environment (as "living solutions" to the physics of life), since the organisms that don't succeed eventually disappear ("unviable solutions to the equations of life").

And so, evolution somehow produced the earliest unicellular organisms: first, archaea and bacteria; later, larger protozoa, algae, and fungi. Unicellular organisms progressively changed the surface and the atmosphere of Earth, and over time they evolved, competed, and cooperated to give rise to multicellular organisms, comprising tissues and organs. At the next levels of complexity, microbes, animals, and plants became part of ecosystems and societies. How can we make sense of this unmanageable complexity in a quantitative way?

ZOOMING IN ON BIOLOGICAL COMPLEXITY: REDUCING BIOLOGY TO ITS BUILDING BLOCKS

As the film *Powers of Ten* reverses and takes us down toward a microscopic view of the universe, we observe the patterns of galaxies dissolving into the simpler motions of stars and planets. As we continue zooming in on the skin of a human and into its cells, life gets reduced to the physical and chemical interactions among DNA, proteins, and other biomolecules. Finally, the complexity of chemistry compresses into the mysteriously geometrical organization of the quantum mechanical atom. Zooming in to smaller and smaller scales of the universe always produces the feeling that perhaps if we knew the workings of the building blocks, and if we were very good at calculating the interactions between them, we could work out the functioning of the whole. This way of reasoning is known as *reductionism*: "The whole can be explained by the interactions of the parts."[6] This is the view that has prevailed to date in much of scientific research in general, and especially in biology.

In biology, this is at least partly because the history of the science has paralleled that of the instruments that made it possible. Knowledge has had to wait for the technology to investigate each of the successive scales of biology. As we know, the universe—

evolution, in fact—has hidden from our eyes the very dimension that makes us possible: the nanoscale.

Instead of nanoscopic vision, evolution gave humans the unrelenting drive to invent instruments to seek further understanding of our environment by looking deeper into it. And for biology, "zooming in" started with the invention of the optical microscope in the seventeenth century,[7] which allowed scientists to discover that plants and animals were made of units (usually) tens of microns in size, the units that Robert Hooke called "cells."[8]

Zooming in further, to figure out what cells were made of, proceeded very slowly, as the next relevant scale was very small indeed: the nanometer (nm). Finally, in the first half of the nineteenth century, chemists revealed the existence of the nanoscale building blocks of life: proteins.[9] By 1900, the notion that proteins were built from smaller units called amino acids was accepted, and all twenty protein-composing essential amino acids had been identified. Soon after, the discovery of X-rays and the technology necessary to produce them made it possible to create one of the most powerful tools for structural analysis known to humankind: *X-ray diffraction.*

The wavelength of X-ray light is so short that it can be diffracted by atoms arranged in three-dimensional lattices (e.g., naturally occurring crystals such as those of sodium chloride, or table salt). From the mathematical analysis of the ordered patterns of light diffracted by crystals, it is possible to figure out the arrangement of atoms within those crystals. The power of X-ray diffraction to reveal the atomic structure of matter revolutionized our understanding of inert materials and, in time, of biological matter as well.

Analyzing X-ray diffraction patterns of biomolecules is particularly challenging because of their complex geometries.

Furthermore, this technique first requires molecules to be arranged in perfect three-dimensional arrays—i.e., crystals—and biological molecules do not lend themselves easily to forming arrays. In fact, making crystals of biomolecules is one of the least well-behaved experiments in modern science, featuring many interconnected variables that cannot be optimized individually. Nonetheless, in the mid-1900s scientists managed to extract the first structures of biomolecules from their X-ray diffractometers, and since then, thousands of proteins and other biomolecules have been crystallized and analyzed successfully with X-rays, thanks to the construction of increasingly powerful X-ray synchrotron facilities worldwide.

In the 1950s, the combined power of the new techniques in chemistry, optical microscopy, and X-ray diffraction led to growing awareness of the immense complexity of life, the intricate organization of millions and millions of molecules crammed into the cell interior: proteins, lipids, sugars, nucleic acids (DNA and RNA), ions, and water. These molecules' coordinated actions and their assembly into cellular organelles and compartments was almost intellectually unimaginable, and certainly beyond the possibility of any rigorous quantitative analysis at the time.

Yet, undeterred, scientists set out to identify, classify, and map the structures and interactions of the cell's nanoscale components, and so the field of molecular cell biology was born. In parallel, the work of the first geneticists had long since established the basic patterns of genetic inheritance, and by the early 1950s their scientific descendants' findings pointed to DNA as the molecule that carried the genes.

A groundbreaking moment was the discovery of the structure of DNA by James Watson and Francis Crick of the University of Cambridge in 1953, famously informed by the data from X-ray diffraction experiments and analysis performed by Rosalind Franklin, her student Raymond Gosling, and Maurice Wilkins at King's College London. The full story of how DNA's structure was unraveled, and by whom, has become one of the best-known dramas of gender politics in modern science; its ramifications never seem to end.[10] The visualization of the "twisted ladder" structure of the DNA molecule, with its four coding units, adenine, cytosine, guanine and thymine (abbreviated A, C, G, and T), as the "rungs" of a double helix, was the culmination of the search for the structure that conferred physical reality on the genes.

The great discovery, however, presented a dangerous temptation. The idea that all the complexity of biology could somehow be sidestepped emerged almost immediately after the first drawing of the double helix. Perhaps the information coded in DNA in simple A, C, G, and T units could explain it all. Surely it was not long before Crick fell for the beauty of the idea and proclaimed the "central dogma" (his words) of the new biology, summarized in his statement: "DNA makes RNA and RNA makes protein." And proteins combine to make the rest. In this description, the four-letter genetic alphabet used by all living organisms is arranged in threes to code for the twenty amino acids, which are concatenated in precise order to form long strings that then fold in specific ways to form working proteins. This picture is correct in broad strokes,[11] but it omits all the details of how a cell uses its DNA in its survival and replication activities, and how, why, and when it reads and translates the code. The dogma in its most simplistic interpretation becomes the explanation of biology: life can be reduced to its building blocks, and biology boiled down to a biochemical algorithm. This simple explanation led to the idea that there would be a unique gene for each protein, a unidirectional flow of information from gene to life, biology as a deterministic process of construction based on a genetic blueprint.[12] Many drew an even simpler conclusion from the dogma: read the sequence of ACGT letters in the genome, and you will know all you need to know to explain life. God "enciphered" all the secrets of life in molecular code and cleverly placed them inside the cell nucleus for us to find and read: nice. Maybe.

Even more appealing, the dogma offered a comfortable view of evolution: it would proceed slowly and gradually, generating new genetic variations (mutations) by pinpoint changes within the genes (random substitution of one nucleic acid letter for another, yielding a different amino acid sequence in the translated protein). The resultant organism would then be selected or rejected by "the survival of the fittest." As Richard Dawkins memorably put it, an organism would be a "survival machine" whose role is to compete, persist, and have sex to transmit "selfish genes" in a hostile environment.

And last but not least, the dogma suggests a very direct and attractive route to medical intervention: find the genes responsible for every disease, then modify and control your rogue genes, and you will be able to fix anything within yourself, diseased . . . or not. I am sure the reader recognizes these ideas, because they are still prevalent in the mainstream media, and they fit nicely with certain political and economic agendas that have struggled to predominate for decades, even centuries. But, as we will see, biology is far more wonderful and meaningfully complicated.

ZOOMING OUT: THE EMERGENCE OF BIOLOGICAL BEHAVIOR OUT OF COMPLEXITY

A large proportion of the science done since its beginnings in the late seventeenth century has concentrated on breaking complex systems into smaller, simpler pieces. This process is reinforced by the complementary development of technology specifically aimed at dissecting and examining matter and natural phenomena at smaller and smaller scales. For many people, this is what science is about: popular representations of scientists often feature a figure in a white coat looking through the lenses of a microscope.

Yet finding out what things are made of is only one aspect of our relation with nature. Humans have always struggled to reconcile the seemingly contradictory properties of our cosmic environment: on the one hand, we manage to find simpler organizational rules in the apparent chaos in which we live; on the other, we are often overwhelmed by complexity. This duality is woven within our soul, mind, and society. Tension between complexity and simplicity lies at the heart of human creativity, in the arts and in our innate urge to create technology (including computing machines) that we use to reshape and explore the world. Human technological power resides precisely in the ability to create simplified models of reality that are translatable into "tools." I would argue that precisely this ability to alter nature through technology produces a delusional sense of superiority that tempts us to explain the world and ourselves in purely reductionist terms. Over the centuries technological progress has often led us to confuse the world with our attempts to dominate it.

Our technological competence entices us to abandon the essential tension between complexity and simplicity that underpins the workings of the cosmos, and invites us to embrace a more comfortable, yet artificial scenario of simpler rules and shapes.

Many of the interpretations and hypotheses that have been put forward following the "central dogma" of biology follow this way of thinking: "Life is assembled from nanoscale building blocks (proteins) following the instructions of DNA." Reductionism is soothing in the face of overwhelming complexity.

But how do we find the right balance? I will present an example from outside the sciences. The Bauhaus movement that so dominates the aesthetics of architecture and design in the modern world (think of Ikea or the iPhone) was born in part to mitigate the economic and social complexity of the interwar period in Germany. Looking for simple, practical shapes and materials that would at the same time make our lives smoother and more comfortable, the Bauhaus pioneers strove for a meaningful coexistence of simplicity and complexity, craft and technology, science and nature. Interweaving arts, design, and architecture, they endeavored to build a path toward survival of the period's nightmarish mix of technological power, destruction, and the chaos of poverty and war.

The Bauhaus did not last very long (although its practitioners continued to spread its influence all over the world, and it is still ubiquitous a hundred years later). The school was closed under pressure from the Nazi regime, which very effectively sought to impose "different strategies" to deal with social and economic complexity. History teaches us again and again that abandoning the complex nature of reality for the temporary comforts of an artificial simplicity can lead at best to stagnation, and at worst to self-destruction. Despite our wishes, nature will never allow us to impose dogmas on her for very long (as the global warming crisis reminds us today).

Simplicity seems to be the necessary scaffold that allows us to fully embrace the fascinating reality we perceive. In the biological sciences, genetics is key, but alone it will never satisfy our thirst for knowledge or our technological and medical ambitions. In reality and in practice, one cannot explain processes such as bacterial infections, electrical waves in the brain, the movements of a shoal of fish, or the human artistic spirit just from relations between genes and proteins.

Yet we must not dismiss our innate capacity to uncover the simpler constituents of reality and to find simple rules. Rather, we must strive to place those constituents in a complete picture of the universe that includes, as well, the complexity of their interactions and manifestations.

By prioritizing reductionism, we have achieved wonders, such as the discoveries of DNA, the nanoscale machinery of life, and the Higgs boson. It is now (again) time to pay more attention to our capacity to feel, intuit, measure, and understand complexity. This, ultimately, is the reason that sciences and technologies are converging in the study of biology in the twenty-first century. If we don't embrace our own biological complexity, technology and medicine will stagnate. The lessons we are learning in doing this, and that I try to summarize in this book, are very deep: nature and history are inviting us to communicate and interact with the world in a deeper way, and hence to enlarge the human capacity to connect fruitfully with the universe.

Reductionism has taught us that creating the instruments to isolate individual constituents of matter is so difficult (imagine the extreme case of the Large Hadron Collider at CERN) in part because in our universe matter is intimately coupled with its environment. Thus it is not surprising that reductionist interpretations alone fail to give a satisfactory explanation of reality. The universe's use of matter, energy, and forces creates structured matter such that the whole behaves differently from the sum of the smaller parts, entangling and merging the properties of constituents and environment—or, as science drily puts it, the behavior of a complex system "emerges."

Let's return to the film *Powers of Ten* (whose authors were deeply influenced by the Bauhaus), this time to take the *zooming out* view of the universe. Here the phenomenon of emergence becomes apparent as patterns emerge from the complex interactions of individual building blocks. For instance, the interactions of billions of stars and planets give rise to galactic swirls which, in turn, gather into clusters and voids. Zooming out, we are forced to acknowledge the cosmic yin / yang duality of complexity and simplicity, the familiar order appearing out of what looked, closer up, like chaos: strong structural fluctuations, unpredictable movements ("dynamics"), and multiple scales of space and time frequently resolve into emergent structure and behavior. Biology is the most fascinating demonstration of the capacity of the universe to create emergent behaviors: shapes that grow complex enough evolve the faculty of "being alive" by coupling their structure to the ability to feel, react to, modify, and even understand the world where they come to life. The fate of a cell is ultimately determined by its intimate entanglement with the environment across the spatial and temporal scales; life interweaves structure, sensing, intelligence, adaptation, and evolution. "Emergence" out of "complexity" has now become a field of study, focused on systems of a nature such that analyzing individual components in isolation cannot explain the behavior of the whole—rendering traditional reductionist approaches to science irrelevant.

Science managed to postpone dealing with the issues of "complexity" until they could not be avoided any longer. Physics spent the early decades of the twentieth century in the reductionist quest to understand individual atoms. The instinct to break matter into its smaller parts gave us arguably one of the most beautiful and mysterious explanations of the workings of the universe at atomic scale: quantum mechanics. But from the depths of quantum theory, physicists once again reached beyond extreme reductionism, reviving the dormant human capacity to intuit enfoldment, hierarchies, and wholes. By interacting with nature though experimentation, quantum physicists began to create physical theory and mathematical models of reality that capture the emergence of simpler behaviors from complexity at smaller scales.

Physics delayed a serious consideration of what are called "collective effects" for nearly three hundred years, but finally, in the second half of the twentieth century, confronted them and managed to explain emergent phenomena observed in solid-state materials. The magnetism of ordinary magnets emerges from the spontaneous collective alignment of the magnetic moment of electrons spinning within their atoms. Similarly, superconductivity and superfluidity emerge from the cooperative flow of electrons and atoms, respectively, at temperatures close to absolute zero (−273°C). In these cases, it is both impossible and pointless to look at the position of each of the electrons in the system to understand the phenomenon. Instead, physics theory based on general physical principles, combined with quantitative experiments to test the theory, has ultimately been able to convey the link between individual electron behavior and the overall collective phenomenon (such as magnetism) without looking at the spin of any particular electron.

Biological complexity is such that its behaviors and manifestations, in most cases, cannot be pinned down to a single definable cause. There are so many active components in organisms, and so many feedback loops between them, that looking for causes in the interactions of the building blocks rarely makes scientific sense. As the reader can see, emergence also has major implications for medicine and clinical interventions: To treat a disease, does it always make sense to look for a single molecular target—the goal of most pharmacology and biochemistry research? And even if that single molecular target were to exist, what is the best way to reach it?

The conclusive proof that biology cannot be explained in a gene-reductionist way came from the very project that had been expected to prove the opposite. In the Human Genome Project, scientists from seven countries collaborated over more than a decade to identify the sequence of bases along all two meters of human DNA. The outcome anticipated for the project was that one gene would be identified for each of the proteins of *Homo sapiens*—the central dogma's laziest interpretation. The number of genes was estimated to be well over 100,000; the precise prediction that was widely advertised was 142,634.

Deliberately, the preliminary results were reported on Charles Darwin's birthday, February 12, 2001. They came as a shock: humans had a mere 21,000 genes—hardly more than the humble roundworm *C. elegans*, which consists of only 959 cells and possesses just over 19,000 genes. Of the two meters of DNA that comprise our genome, only 8 percent to 15 percent is occupied by "genes," sequences that code for proteins.

Human complexity could not be generated, after all, by the central dogma's most reductionist approach. The Human Genome Project demonstrated that in order to create more than 100,000 proteins from the information encoded in 20,000 genes, the DNA code must somehow be manipulated and multipurposed; the cell must be choosing, cutting, and pasting pieces of genes in response to its interactions with the environment. We now know how this happens. A single gene is composed of coding (exon) and noncoding (intron) segments; the instructions to assemble a protein consist of the exons spliced together after removing the introns (a process which takes place during transcription from DNA into

RNA). If some exons are omitted or arranged in a different order, then several proteins can be created from the same gene. Crucially, the way in which this is done must be linked to many nongenetic (epigenetic) factors, which include molecular labels inserted on the genomic DNA, but also the arrangement of information within the genome and the structure and physics of DNA inside the nucleus.

The results of the Human Genome Project force us to abandon oversimplifying models of life based on a linear code of DNA, and invite us to come up with more-encompassing scenarios by considering the possibility that very complexly arranged biological matter has the capacity to compute solutions for survival. Let's try to imagine how the cell would use the information encoded in DNA to react to signals coming from the environment. The two meters of DNA confined in the tiny (5-micron) cell nucleus should be able to work as a kind of "living computer" that responds to chemical and physical information incoming from the rest of the cell by tuning cell performance to current conditions or challenges. This could be done using information in stretches of DNA that are strategically and dynamically placed to react to the incoming signals. The solutions would consist partly in expressing proteins that facilitate certain cellular "behaviors," such as differentiation or migration, to promote cell survival.

To sum up, evolution creates biological matter that is able to encode, in the hierarchical complexity of its structure, the capacity to read and recognize the environment, and to react to it by computing optimum survival strategies. Biological structures act, then, as the algorithm itself, able to learn, adapt, and evolve in real time. In this way, in life, order and macroscopic behavior emerge from the use of the sophisticated computing power created by the physics of the universe. This involves the transfer of information and energy, and the creation and rearrangement of ordered yet complexly interconnected structures (powered by what physicists call non-equilibrium thermodynamics, as first realized by Erwin Schrödinger in his influential 1944 book, *What Is Life?*). This picture is much more plausible and appealing than a rigid string of "selfish" DNA code, and points toward the physical origins of intelligence itself. It also hints at how life reconciles complexity and simplicity by coupling perception and computation in the dynamic arrangements of its molecular structure.

After the Human Genome Project, the emergent biology of the cell came into the spotlight, and we were forced to move on from biomolecular reductionism. The cell can also sense cues that are not molecular, such as mechanical and electrical signals and vibrations, or changes in the temperature or chemistry of the environment, and can use DNA to react to those cues in ways that may be genetic or nongenetic. For example, a sharp tug at the cell surface could be transmitted all the way to the cell nucleus, deforming it, and this could in principle trigger a mechanical rearrangement of the DNA inside the nucleus that might result in the increased expression or repression of specific proteins. The shape of the nucleus, its mechanics, its connections to the rest of the cell, and the arrangement of DNA inside the nucleus all now became part of the living computing power of the cell. Going forward, new ways of interacting with the genome from outside the cell would have to be incorporated into biological research and the development of effective medical treatments.

The acceptance that life emerges from extremely complex collective effects has deep implications for the way we study and understand life. The first is that we can happily and confidently abandon the harmful idea that every aspect of our character and

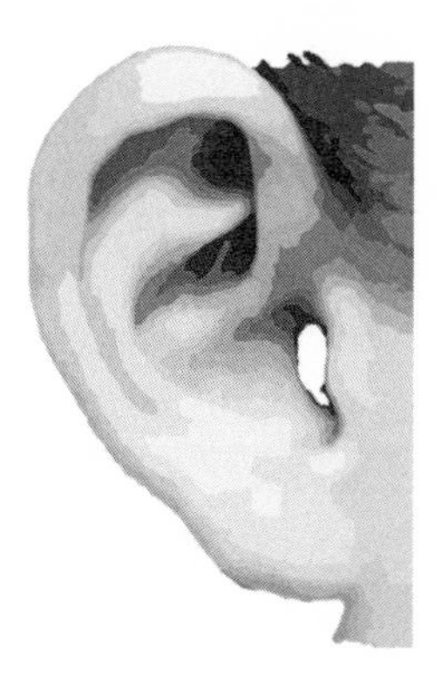
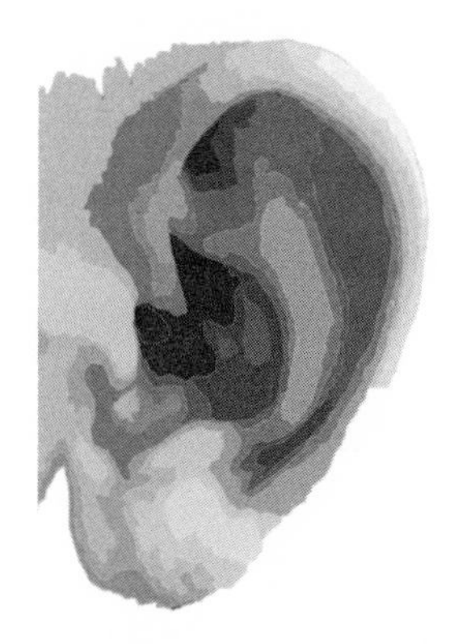

Attached earlobe Dangling earlobe

Figure 1.1. Do you have dangling or attached earlobes? Although this was once promoted as textbook example of a trait controlled by a single gene, recent research has demonstrated that, in fact, hundreds of genes actually influence this inconsequential characteristic.

physical complexion can be ascribed to a particular gene. Our destiny is not written in our genes, as most of us already knew.

This wider view also has profound repercussions for medicine, and it means that the biotechnology industry's rush to patent genes, hoping to secure commercial rights to them, was fortunately unfounded. Fixing aberrant genes will not cure every single human illness. Although some illnesses can be attributed to the expression or silencing of a particular faulty gene, this assumption cannot be made about diseases or traits in general. For example, the peculiar penchant of human earlobes to either attach or dangle at the base was once propounded as textbook example of a trait that is controlled by a single gene. However, recent research has provided detailed evidence that hundreds of genes actually influence this apparently inconsequential human characteristic.[13]

It's not just earlobes. It appears that in the determination of human height, and in the risk and development of schizophrenia, rheumatoid arthritis, and Crohn's disease, all genes in the genome might contribute to the actual condition. A new "omnigenic" model has in fact been proposed, postulating that complex traits and diseases may be modulated by all the genes that are active.[14]

This prompts the question: Is just looking for the genes involved in a biological process or form the most sensible way to explain it? Evidence seems to point to larger organizational and functional principles that direct and combine the expression of genes, since "everything in a cell is connected."[15]

Recent research has also shown that the cells in our bodies are not genetically homogeneous. Our bodies seem to be a kind of genetic pastiche. Contrary to the widespread belief that each of our cells contains a replica of the DNA that started our life as the fertilized egg implanted in our mother's womb, mutations, copying errors, and editing errors start soon after the first divisions of the embryonic cells. By the time we reach adulthood, the DNA of our cells is sprinkled with mistakes: missing, rearranged, and repeated information, and even the loss of whole chromosomes. Most research has assumed that anomalous DNA leads to disease, but this picture is slowly changing: genetic diversity seems to be useful, as, for example, in the liver's ability to regenerate itself when damaged by diseases such as cancer, hepatitis, or infection. The genetic diversity of liver cells probably confers on them different abilities to survive and regenerate tissue after damage, while identical cells could potentially be wiped out by a particularly effective infection. Genetic diversity might be the key to creating a cell population that is more resilient. A certain degree of genetic variability has also been found in the brain, and although the results are still preliminary and controversial, some scientists speculate that this genetic variability might be somehow related to human individuality and adaptability, our capacity to adjust to the different circumstances of our lives. The brain's genetic diversity "is compelling because it reflects a truth about human nature, that our vulnerabilities and our resilience are all bound up together, two sides of one coin."[16]

In 2018, scientists attempted to disentangle the interactions of genes that are necessary to keep yeast alive. Yeast is a unicellular organism that is present in our daily lives in bread or beer making, but also in the lab, as model system of eukaryotic biology. They performed hundreds of thousands of experiments in which they deleted three genes of the yeast genome at a time. Their results showed that almost all genes effects are in fact interconnected and that deleting one single gene almost always has consequences for the health or survival of yeast cells. There seem to be no "essential" genes for yeast survival, as it needs almost all of them to be alive. This study also has broad implications for gene editing in human genetic interventions in the future.[17]

In the twenty-first century, science has no other intellectually viable route than to embrace life as "a symphonic interplay between genes, cells, organs, body, and environment."[18] In the next sections, I will show how the deployment of nanotechnology and the participation of physical and mathematical scientists in biological research are contributing to advances in this direction.

USING THE TOOLS OF NANOTECHNOLOGY TO INVESTIGATE BIOLOGY

Arguably, one of the reasons why reductionist hypotheses were so comfortable for much of the biological research community was the fact that one of their main tools, X-ray diffraction, requires that molecules (DNA fragments, proteins) be arranged in a crystalline array, able to diffract X-ray radiation in a pattern strong and clear enough to allow mathematical analysis. Trying to crystallize biomolecules is an unimaginably difficult task that can take decades to carry out for a particular molecule—and it turns them into blocks of an artificial material that is studied in dry, unbiological diffractometers. For the powerful and well-funded scientists in the field of protein crystallography, proteins and other biomolecules became inert crystals removed from life. Thus it is unsurprising that reductionism became the status quo.

The science of nanotechnology was born with the arrival of tools able to visualize and interact with matter at the nanometer scale in a direct way, without the need for crystallization or X-rays. This happened in the 1980s, when the tool that marked the "before and after" moment finally materialized: the scanning tunneling microscope, or STM. The extremely ingenious inventors of the STM created a microscope that sidestepped the limitations of conventional light-based microscopy by interacting with nanostructures, and even individual atoms, using a very sharp—atom-thin!—tip. By scanning the tip over the sample in a very controlled way and mapping the local interactions of the tip with the exterior of the sample, an image could be created that delineated the surface with atomic precision. With the STM, individual atoms (and, later, their electronic interior) at last became "visible," or more accurately, perceivable.

But beyond imaging with unprecedented accuracy using a relatively simple, cheap tool, the STM had drastically new capacities: it was able to *pick up and arrange* atoms one by one (see fig. 1.2). Importantly for the story I am telling here, the STM could also map collective phenomena arising from the complex interactions of atoms (such as the standing wave of electrons emerg-

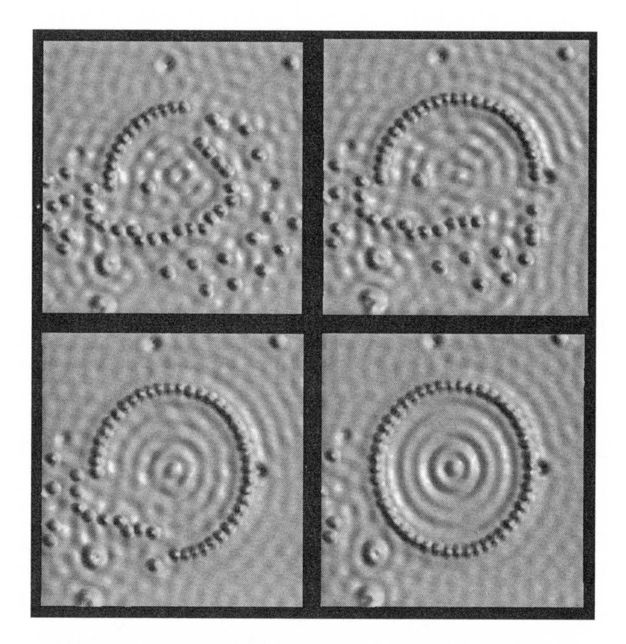

Figure 1.2. STM image of iron atoms (the little spheres) on a copper surface (the smooth bumps are copper atoms arranged in a crystal). The STM tip is used to arrange the iron atoms in a circle, one by one. This is the proof of principle that humans can construct objects with atomic precision. When the atomic circle is completed, the concentric ripple of a standing wave arising from the collective movements of electrons in copper shows one of the most beautiful visualizations of collective effects in the quantum world of atomic matter: electronic waves confined in a "quantum corral." *Source:* These images were taken by Don Eigler at the IBM Almaden Research Center in San Jose, California, in 1991 (Image originally created by IBM Corporation)

ing inside the "quantum corral" visible in the fourth panel of fig. 1.2).

This image is not only *a first*, a proof of principle of human ability to manipulate the ultimate constituents of matter; perhaps more importantly, it merges the reductionism of single-atom building with the visualization of a complex phenomenon emerging from the collective motions of electrons. With the STM, a fundamentally different level of human interaction with matter had started.

The STM cannot be easily deployed in biological environments because it requires a voltage to be applied between the tip and the sample immersed in liquid, which is not practical in most cases, as the voltage in the tip can produce chemical reactions that are difficult to avoid or control. A close relative of the STM, however—the atomic force microscope, or AFM—can easily be adapted to produce images of single proteins or single DNA molecules, whole cells, or cell components. Crucially, the sharp tip of the AFM can be immersed in biological fluids, and is able to map biological surfaces by controlling the force between the tip and the sample with an accuracy in the range of tens of piconewtons. (A piconewton is 10^{-10} newtons; a newton represents more or less the force of the weight of an apple held in one hand at sea level on Earth. A piconewton represents very little force indeed—more or less as much force as biomolecules in solution "feel" when warm water molecules collide with them.)

The AFM is able to create maps of the surface of biomolecules with sub-nanometer accuracy, and, crucially, it can be used to selectively manipulate the sample. In plate 2, I show a DNA image taken with an AFM in my lab, showing the details of the double helix. The AFM user is able not only to visualize single molecules and their structural details, but also to use the device to push and pull those molecules, to inject electrons into them, or to put them in contact with other molecules (for the purpose of measuring the key quantities that can be used to create mathematical models of their function).

Nanotechnology tools like the AFM and STM link the macroscopic world with the nanoscopic world in a direct way. Nanotools create, in fact, *a new sense* for the humans that operate them, since we can use them to *feel* the chemistry and physics of the microscopic world of biological cells and biomolecules with nanoscopic sensing capability, a bit like a protein does.

With the new tools of nanotech, biology moved closer to the realm of physics, because fundamental quantitative questions about biology could now be addressed: Why is the nanoscale special for the emergence of life's molecules? How do proteins use water and ions and temperature to function? How do their collective actions emerge into cellular behaviors?

A beautiful example of how the AFM can be used to understand life at the molecular level is how the workings of the ubiquitous *myosin* protein are being investigated.

OBSERVING THE FUNCTION OF BIOMOLECULES: A PROTEIN PERFORMING NANO-WALKS

Myosins are proteins that are present in different forms and variations in all the cells of our body. They are the nanoscale machinery underlying the contraction of our muscles; in hearing, they are responsible for key processes in the inner ear. Myosins are also in charge of moving cargo around cells, and are instrumental in activating some of the cell's skeleton (*cytoskeleton*) filaments made of the protein *actin*—thus contributing to cellular contractions in vital processes such as cell division and cell crawling. If a biological process requires cellular contraction, it is likely that myosins are involved. Virtually all cells of eukaryotes (including both animals and plants) contain myosins.

Myosins have a surprising way of carrying out their work: they can "walk" on molecular actin tracks. Myosins are usually elongated molecules with two "heads," a "neck," and a "tail." A part of the head acts like "feet" that bind to the actin track and step along. The force generated by the head domain is transduced by the neck, which works as a lever arm, and the tail is attached to the cargo (in myosins that carry loads as they walk). How does this protein achieve this remarkable activity? First of all, myosins need energy, and secondly, they need a clever way to utilize it to move forward.

Proteins such as myosin have two fundamental ways of getting energy. The first is the thermal energy of the cell. What we feel and measure as "temperature" means "movement" at the molecular scale. When the temperature is above absolute zero (–279°C), molecules and atoms move; the higher the temperature, the faster the molecules are moving. Our bodies take temperature very seriously because our biomolecular building blocks are optimized to work at a constant temperature, around 37 degrees Celsius. At this temperature, water molecules (which are rather small—they comprise just three atoms) can move very fast (around 600 meters per second!). In water, a protein like myosin is constantly being bombarded by water molecules, and although it is orders of magnitude bigger, it can still register the collisions. In fact, proteins are subject to a continual shaking movement that is produced by the collisions of water molecules (also known as *Brownian* or *thermal motion*). The mechanical design of the myosin molecule is such that it can use the energy and direct the turbulence of the shaking movements to drive a processive "hand-over-hand" movement—detaching its trailing head and moving it forward to reattach to the track. (This is a bit akin to the structural design of turbines that makes them able to rotate when wind or water passes through them.)[19]

Although thermal energy seems to be enough for getting the molecule to step forward, myosin uses another trick: the heads can bind ATP (*adenosine triphosphate*, the biological molecule that acts as the currency of energy) to accelerate the movement, thus harnessing chemistry in the operation as well. When ATP binds to a special binding site in the head, it releases a phosphate molecule and around 10^{-19} Joules of energy. The molecule uses this energy to detach the trailing head from the track; once it is detached, the forward swing of the new leading head and the resulting forward step occur spontaneously.

Studies of myosin molecules started in the 1960s, but only with the advent of nanotechnology techniques that could directly address single molecules did it become possible to probe the mechanisms of action of individual myosin molecules. In the 1990s, *optical tweezers*, which can trap the two ends of an actin track between two microscopic beads controlled by lasers, were used to study the forces that single myosin molecules exert as they step along. Years and hundreds of researchers' efforts have produced the picture that I just sketchily explained.

The arrival of high-speed AFM has greatly facilitated the detailed study of this "linear molecular motor," for with it the stepping motion can be observed directly, in warm salty water. In 2010 the first movies of myosin molecules walking on actin tracks were published.[20] I reproduce in plate 3 the amazing details of their walking movements. It is astonishing that these tiny molecules can produce such activity. I invite the reader to admire the high-speed AFM videos available online, and to reflect on the fascinating nanoscale molecular action that underpins so many of our interactions with the external world, including our own walking.

These experiments reveal that molecular biology is much more than chemistry. The key to biomolecular function (and to the emergence of life) is that, at the nanoscale, the interactions of structure—mechanical and electrostatic design—with temperature, water, ions, and chemical reactions can produce deterministic movements. At the nanoscale, physics and chemistry meet and couple to maximize living processes' efficiency; mechanics and movement are the fundamentals of biological activity.

The movies of myosin beautifully exemplify the capacity of the tools of nanotech to cross the temporal and spatial scales and to generate quantitative data, essential for developing models that can explain and even predict the behavior of biomolecules. The detailed biophysical studies performed in the last thirty years have been incorporated into increasingly sophisticated mathematical models and simulations of myosin, comprising information about its mechanical design and electrostatics.

Similar models and experiments are being done for other crucial molecular motors, such as ATP synthase, a protein enzyme that can function as a rotary motor, using mechanical rotation to catalyze the chemical reaction that produces the "energy currency" ATP. I have contributed to this field by studying the mechanics and movements of the archetypal membrane protein bacteriorhodopsin.

CELLULAR BEHAVIOR ON MULTIPLE SCALES

The new tools coming from physics labs are allowing scientists not only to interrogate single biomolecules, but also to investigate how the concerted actions of molecules give rise to emergent cellular properties, enabling cells to inspect, adapt to, and modify their environment at other scales.

Advances in optical microscopy (the so-called super-resolution microscopies) are making it possible to identify individual molecules and their movements inside living cells. High-speed optical microsocopies (especially advanced light-sheet microscopes) are showing us the mind-blowing details of the collective movements of cells and their structures, during embryonic development, for example, and even inside living organisms. Beyond showing details and dynamical changes, some of these microscopies (such as raman or infrared microscopy) and AFM are also capable of chemical identification and mapping of the mechanical or electrical properties of cells and tissues.

Apart from looking at biology with powerful microscopes, a new way of interrogating cells has evolved into a distinct field in recent years. To investigate the complex ways cells interact with their environment at different scales, biophysical scientists have learned from nanotechnologists how to fabricate structures with physical and chemical features displayed in micro- and nanopatterns. By varying the composition and structure of the patterns and observing how cells attach to and interact with them, these researchers are elucidating the mechanisms that cells use to sense, modify, and adapt to their environment. This is analogous to fabricating "soft toys" with cleverly patterned nanofeatures and giving them to cells to "play with," then observing that play for insight into cellular functions—a kind of "quantitative cell psychology."

The origins of these ideas can be traced back to the 1990s, when scientists started to investigate how cells reacted to microstructured materials.[21] Most cells adhere and adapt to the patterns and shapes, changing their movement and behavior according to the topography they encounter. I show an example in plate 4, again from one of my own experiments. Presenting patterns to cells has allowed scientists to better understand cellular behaviors such as adhesion and locomotion, and has pushed research toward questions about the role of mechanical forces in cell function.[22]

The role of mechanoforces in biology had already been explored by the powerful intuition of the Scottish scientist D'Arcy

Wentworth Thompson one hundred years ago; however, the predominance of reductionist molecular cell biology in the twentieth century pushed mechanics out of most biological research. At the turn of the twenty-first century, the ideas that Thompson outlined in his classic book *On Growth and Form* (1917) were starting to be revisited. Modern research is confirming what D'Arcy's intuitions anticipated: forces, scales, and mechanics are necessary to explain biology and its behaviors. A new generation of scientists and engineers, armed with powerful new nano- and micro-instruments, is now ready to bring physics to biological problems in which forces and mechanics (together with chemical and electrical signals) play a key role, such as developmental biology, stem-cell differentiation, and, crucially, the new field of tissue engineering, to be explored further in chapter 5.

Using this approach, research has demonstrated that tissue-forming cells not only feel the chemistry of their surroundings, they mechanically sense their environment, too. Presented with an external material, the first thing cells do is attach their nanoscale "hands" (cell-surface and transmembrane proteins such as *integrins*) to it. To investigate how they do it, experimenters expose cells to molecular sticky patterns—sticky molecules displayed at varying distances and geometries on substrates of different stiffnesses. Integrins particularly like to stick to molecules that present a sequence of amino acids specific to certain proteins, such as collagen and hyaluronic acid, that characterize the *extracellular matrix* in which cells of most living tissues are embedded. The cells' attachment and ultimate behavior depend critically on the substrate's rigidity. If it is too hard, cells detach[23] (which can explain, for instance, the failure of hip implants). But if they stay attached, they tug on it and activate genes in the cell nucleus to adapt to it. This is a crucial finding, because it demonstrates that mechanical forces can trigger the reading and transcription of the genome and its translation into proteins. Cells can adapt their composition and structure to their mechanical environment.

This exemplifies how cells use structures at different scales to sense and respond to the environment. They attach at the nanoscale through proteins, enabling pulling and cytoskeletal deformation at the nano- and microscales, which is then fed back as information into the nanoscale of DNA. This facilitates a fast connection between the exterior of the cell and the nanoscale machinery inside it that is able to respond to external changes. This *mechanotransduction* pathway for cellular behavior is a radically different scenario from the traditionally reductionist biochemical vision of molecular cell biology and genetics. The new view recognizes the cell as a very clever structure that can sense its environment in real time via millions of nanoscale signals, translate those into its nucleus, and adapt to the environment by activating genes. It is a given that the mechanical properties of the environment are fundamental to the process, because much of the communication within the cell is mechanical, and much of the environment exerts mechanical forces on the cell.

HOW DO WHOLE CELLS RESPOND TO FORCES AND THE MECHANICAL ENVIRONMENT?

In 2006, a game-changing experiment was reported:[24] a number of mesenchymal stem cells (a type of stem cell produced mainly in bone marrow that can *differentiate* or specialize into bone, cartilage, muscle, or fat cells in our bodies) were divided between three culture plates coated with flat polymer matrices of varying degrees of softness. The first substrate was very soft, around 1 kilo-Pascal, mimicking the mechanical properties of brain tissue.[25]

The second had the softness of muscle (about 10 kiloPascals), and the third had a stiffness closer to that of bone, around 30 kiloPascals. After several days on their respective matrices, the cells started to behave differently in each of the three culture plates: more like neurons on the softest substrate, resembling muscle cells on the intermediate one, and more akin to bone cells on the stiffest one. It appears that by attaching to the substrate and pulling against it, the cells feel its mechanical properties, and then they react and differentiate, producing proteins that help the cell transmute into a cell that would be appropriate to the sensed environment. The cells try to match the stiffness of the material they grow on, and in doing so, they turn into particular internally programmed types. In other words, a differentiated cell can be distinguished by its special mechanical character, not only by its molecular biomarkers or gene expression patterns. This experiment has profound scientific, medical and philosophical consequences: it demonstrates that cells have "mechanical identities," and that human-made materials can be used to steer their behavior into those identities.

In 2010, this experiment was recapitulated in a more lifelike environment. Cells were embedded in a three-dimensional network of biopolymers constructed to mimic the nanoscale mechanical and chemical properties of the extracellular matrix of real tissues. The results showed again that stem cells differentiated according to the mechanical properties of the material. Furthermore, the matrix stiffness also regulated the organization of integrin (the protein acting as the cell's "sticky hands") at the cell surface. To assess the environment, cells need to adhere and pull against it with nanometer "hands" to feel its stiffness and its three-dimensional topography. Then the cell responds by adapting to achieve the most "comfortable" equilibrium state. As it adapts, the cell also remodels the environment to make it more favorable for living and reproducing.

Cells require the right topography, and particularly the right mechanical environment, to behave healthily.[26] In chapter 5 I will explore how these ideas are being used in regenerative medicine to create or repair tissues and organs.

TRANSLATING MECHANICS INTO BIOLOGY

How do cells translate mechanics into the production of proteins and cell differentiation? In other words, what actually happens in cellular mechanotransduction?[27] We still don't know. Molecular cell biologists are looking, as they do, for individual molecular messengers—proteins that trigger complex cascades of molecular activity that eventually cause the expression of genes. But physicists and engineers are looking for other possibilities. One of the pioneers is Donald Ingber, an American bioengineer and the founding director of the visionary Wyss Institute for Biologically Inspired Engineering at Harvard University. He has put forward the possibility that when a cell sticks to and pulls from the substrate to gauge it mechanically, its nucleus gets deformed as a result. The cell has a markedly mechanical architecture: it is crisscrossed with fibrillar proteins acting as interconnected nanocables, mainly actin filaments, microtubules, and intermediate filaments. These form the *cytoskeleton*, which Ingber first identified as a natural tensegrity structure[28] (see fig. 1.3)—one that maintains its shape and stability through tensional force.

Suggestively, this intricate mesh of nanocables is connected both to the external nano "hands" of the cell (the integrins and other proteins that attach to external materials) and to the cell nucleus that contains the two meters of genomic DNA. It is likely

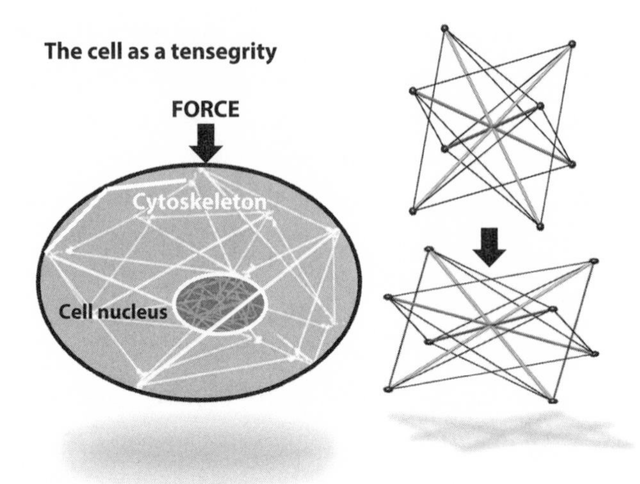

Figure 1.3. It has been proposed that the cell behaves like a tensegrity structure, in that when it is deformed by external forces its whole structure responds, creating forces that are transmitted to the cell nucleus, and thereby affecting the way the cell reads and translates the genome.

that an external mechanical pull can be transmitted through the mechanical structure and result in a deformation of the nuclear DNA. The extremely tight yet highly structured and hierarchical packing of DNA inside the nucleus would ensure that the force can be selectively transmitted to facilitate or hinder the expression of specific genes (fig. 1.4 and plate 5).

To achieve this the cell must have very clever pathways of force transmission. These likely include the number and distribution of external cellular "hands" and a precise sensitivity to the actual value of the force that is transmitted. The cytoskeleton's organization and the cell-surface positioning of integrin and other adhesive proteins must ensure that an optimal intensity and refinement of mechanical signals reaches the nucleus so

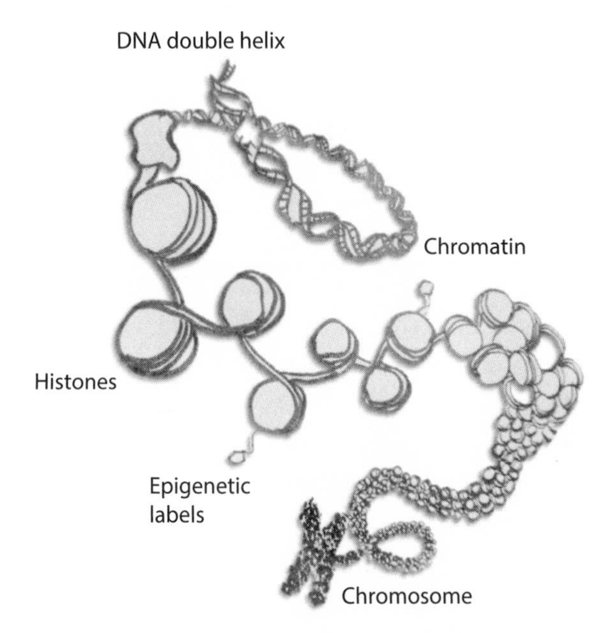

Figure 1.4. Cartoon of the hierarchical assembly of DNA inside the X chromosome. DNA is wrapped around histone proteins that are nicely folded into chromatin fibers, which in turn are then packed in specific positions along the chromosomes. Some histones, and DNA itself, may have *epigenetic* labels attached: active molecules that mark, for example, whether a gene should be activated or not, or that modify chromatin's local structure to "open" or "close" access to particular genes.

the cell can "understand" the demands of its environment. It is conceivable that mechanotransduction can even activate combinations of genes: some pulls and forces might expose areas of DNA whose mechanical and electrostatic properties make them susceptible to be read and translated at those particular force ranges. (See plate 2 for an AFM image showing that the deformation of DNA under tension can unravel the double helix.[29]) The persuasiveness of some of these ideas is increasing with each piece of new research, but so far they remain mostly unproven, and ingenious experimentalists are finding ways to investigate them.

Mechanotransduction also proposes a more physical formulation of adaptation and evolution than does the "central dogma" alone. The way the genome, packed inside the nucleus, moves and deforms in response to external forces becomes an important way that the information coded in the genes can be read and combined by the cell, using both chemical and physical mechanisms.

These discoveries are putting cell mechanics at the forefront of biology. Techniques are being developed to measure the mechanical properties of cells, and to probe their remarkable abilities: to produce mechanical signals or even mechanical waves, to transmit or dampen mechanical cues, and to respond to them. Cell biology is progressively becoming the most sophisticated branch of mechanical engineering and materials science. Of course, this is a radically complex materials science, in part because, in biology, mechanics is always coupled somehow with biochemistry and electricity, and in part because of the multiple, interlocking feedback loops that biological systems keep going in order to keep living. For a living structure, the interconversion of energy is fundamental: ultimately, we must produce movement to interact with our environment, and we must fuel that movement with energy extracted from the food we eat. This requires that the energy obtained from a chemical reaction can be translated into movement at the nano-, micro-, and macroscales. Conversely, mechanical movement—a change in shape, a deformation—can facilitate a chemical reaction or direct a chain of chemical reactions: enzymes such as ATP synthase, for example, use rotation to continuously catalyze chemical reactions. A change in electrical properties or voltage in a biological structure can modulate the chemistry or the transmission of mechanical forces. Biology does not distinguish between the domains of science; it uses them all. In other words, mechanics alone will not describe life, but neither will biochemistry or genetics alone.

As the reader can imagine, the mechanical properties of living cells and tissues are not easy to measure, especially at the scale of proteins and subcellular structures. Mechanical properties can be discerned in several ways, but the most direct and accurate is to deform the sample that we want to study. By pushing or tapping a material with precise control at different depths and speeds, one can learn—from how much and how quickly it is deformed by varied forces, and how fast it recovers—how stiff, elastic, or viscous it is.

The AFM, since it is a mechanical microscope, can be used not only to image cells and biomolecules with high precision, as we saw before, but also to "tap" cells and learn about their elasticity, their viscosity, and the time it takes them to react to forces.

The unique accuracy of the AFM's nanometer-scale "finger" tip is being used in labs around the world to measure these mechanical properties in biologically relevant environments, and AFM companies are refining the technology for this purpose. I am a part of this scientific endeavor, collaborating with others to create methods that can map the mechanical properties of living cells with nanometer accuracy using the AFM.[30]

Quantitative measurement of the mechanical properties of cells is important for many reasons; one is that physicists, mathematicians, and engineers can use the data to simulate in their computers how cells will behave—say, during growth—given their particular mechanical characteristics. For example, my lab currently "feeds" AFM data on cell mechanics to engineers who are making computer models of such processes as the growth and formation of plants, or the transmission of mechanical signals in the brain.

Mechanics and forces are fundamental for the way cells feel their environment, and this is true not only for cells that form tissues, but for all living cells—for another vital example, the cells of our immune system. These cells are responsible for detecting anything from outside the body, organic or inorganic, that has penetrated the body or attached to one of its surfaces (or something misbehaving inside the body, such as a cancer cell), and getting rid of it if it is dangerous. Immune cells cannot afford to make mistakes; if they misidentify an integral part of our body as an invader, the consequences are the self-destructive effects of autoimmune disease.

Combinations of research have, for example, shown that immune cells can recognize bacteria by nanopatterns in their external topography. Bacteria are coated with sugar molecules (carbohydrates), and T cells and macrophages can identify them both by the type of sugar molecules and by their arrangement on the bacterial surface. It appears that besides attaching to and "tasting" the sugars, the immune cell also deforms the bacterium by pulling on the sugars, and uses this information to decide whether to destroy the bacterium or not. The immune system looks for the right combination of "flavor" and "texture" in its microbial prey to establish whether they are pathogens or not. These findings hold promise for understanding not only how to better combat infections, but how to fight cancer as well. One of the tricks that tumor cells use to evade the action of the immune system is to mask themselves and confuse detection mechanisms with clever combinations of "flavor" and "texture" at their surface. Retraining immune cells to detect cancer cells and destroy them is currently a very active area of biomedical research, as will be detailed further in chapter 3.

BRIDGING SCALES WITH MECHANICAL AND ELECTRICAL SIGNALS

A characteristic of mechanical signals is that they can travel relatively long distances. We know this because we use them to communicate with each other. Sound waves are mechanical signals propagated through air (and water) that our ears can detect and our vocal folds can produce, using the wonderful nanomachinery in our inner ear and vocal folds. Feeling sound is not confined to specialized anatomy, however. In fact, all our cells are sensitive to mechanical vibrations. To mention a celebrated example, the total deafness of the percussionist Evelyn Glennie has not precluded her becoming a world-class musician, as she taught herself to hear with parts of her body other than her ears.

All evidence so far indicates that cells use forces and mechanical signals to transmit information across the nanometer and micron scales to reach the milliliter and even the centimeter scale. Cells have structures that enable them to amplify nanoscale forces and vibrations, and to generate and propagate mechanical signals—even waves and oscillations—through tissues. Indeed, new, ultrafast imaging microscopies have begun to show the microscopic details of processes such as embryonic development, in which cellular vibrations are coordinated and synchronized, giving rise to patterning and mechanical waves traveling through the embryo. The movies made with the new microscopes indicate the possibility that mechanical waves are being used, together with chemical and electrical signals, to harmonize the development of tissues and structures during embryonic development.

The classic textbook example of long-distance multiscale propagation of signals in biology is the action potential of neurons that transmit information, such as that collected by our senses, to the brain to be processed through the networks they form in the nervous system. The action potential is an electrical signal that propagates along the surface of a neuron and can travel down its axon for quite a long distance. Axons can reach meters in length in the some of the longest neurons, such as those found in giraffes and whales. The brain can also send signals, e.g., to our muscles to interact with our environment, or to viscera or glands to control bodily functions, using action potentials. The measurement and mathematical modeling of the action potential (the propagation of the electrical spike) in the 1950s was one of the earliest successes of quantitative approaches in biology.[31]

Both mechanical and electrical signals are key to embryonic development, cancer, and wound healing. Mechanical and electrical signaling in biology are very complex, and our understanding of them is still very basic, but they are good examples of how collective phenomena, arising from the complex interplay of biomolecules in living organisms, connect the scales of biology from the nanometer scale of atoms to the meter scale of our bodies. These phenomena and their mechanisms are finally the subject of intensive quantitative biological research by several disciplines.

BIOELECTRICITY PROGRAMS ORGANS' ACTIVITY

The picture emerging from the combined efforts of different approaches is that cell behavior is regulated by chemical, mechanical, and electrical signals that are often coupled (it is now confirmed, for example, that the propagation of the action potential is accompanied by a simultaneous mechanical wave, whose role remains unknown). Together with the efforts of the biophysical community to understand the role of mechanical forces in creating biological shapes, Michael Levin's work revisiting the crucial functions of electric fields in tissue regeneration and wound healing, embryonic development, and cancer has attracted much recent attention. This work carries an important message that cannot be overlooked: to create bodies, organs, and "biological shape" in general, organisms use long-range signals that are not specified in the genome. Working with tadpoles, Levin and collaborators have shown that the shape of the embryo is modulated by electrical signals generated in its brain even before that organ is fully formed. Striking experiments from his lab have shown that electric fields can make a tadpole grow an eye in the gut, or a worm develop two heads. Levin thinks that electrical signals in bodies create bioelectric circuits that encode for anatomical patterns. Understanding the role electric fields play during development might be key to future regenerative medicine, cancer biology, and bioengineering strategies. Levin speculates that in the near future, using advanced techniques and computer science (or, more precisely, *computational neuroscience*), we will perhaps start to understand the *semantics* of these bioelectric circuits, and finally crack the bioelectric code of development and regeneration.

Although the field of quantitative biology is in its infancy, as it takes on the daunting task of looking for the links between biological behaviors and complex underlying structures and processes, a clear research strategy underlies most studies. Advanced microscopies are combined with biochemistry and genetics; nanostructured materials and the tools of nanotechnology are used to explore the different scales of cellular and tissue behavior. Finally, physics, mathematics, and computer science are being deployed to construct increasingly accurate models of biology that endeavor to embrace all of its complexity, from the molecular level to the cellular to the organismic.

HIERARCHICAL BIOLOGY, HIERARCHICAL BRAIN . . . AND MIND

Perhaps the most evidently emergent characteristics of living organisms are intelligence and consciousness. These have traditionally been beyond the ambitions of most biological sciences, since reductionist approaches can hardly begin to tackle such questions. But recent research on machine learning and AI has proven that quantitative approaches and mathematical modelling can bring very profound insights into living organisms' ability to recognize patterns in their environment, learn from them, and even extract abstract concepts.

In March 2016, AlphaGo, a computer program developed by Google's DeepMind, defeated Lee Sedol, one of the world's top players of Go, in a five-game match. This was an incredible feat for AI. AlphaGo used an artificial deep neural network to learn by training with humans and computers to play and to win at Go. These days, deep neural networks overshadow humans at recognizing faces—something that was thought impossible only a few years ago—and are becoming increasingly better translators between languages.

The key to the success of newfangled deep learning algorithms is that they are inspired by the architecture of the human central nervous system; they are constructed with nodes that are connected together to form something resembling a biological neural network. They are built in hierarchies, in which neural networks or parts of networks become components of larger systems. But, despite their success, they are still "black boxes" to their creators. Hidden in their complex, layered connectivity resides the foundation of their learning ability, but we don't know how they do it.

An interesting hypothesis to explain why these AI systems work so well turns on a key property of the universe that they exploit, as neural networks do: the hierarchy of sizes and structures demonstrated in the film that inspires this chapter, *Powers of Ten*. The universe is hierarchically constructed, and complex structures are often formed through a sequence of simpler steps. The layered structure of artificial deep neural networks enables them to mimic that hierarchy of causality in the sequence of their calculations, and this makes the calculation easier. Furthermore, our universe seems to be governed by a tiny subset of all possible functions, and these functions have simple properties. The structures of the universe are created by these simple functions. This fact can be exploited by neural networks when, for example, they are trying to identify a cat's face. In this case, they do not need to approximate an infinitude of possible mathematical functions to identify the cat, and so the calculation can be done in a short time. When a phenomenon has a hierarchical structure, as does the picture of the cat, layered biomimetic neural networks are much better at analyzing and modeling it than computer algorithms that don't have that layered structure.[32]

Bioinspiration for computing is not only being practiced by mathematics researchers. Experimental scientists apparently "bored with reductionism . . . tired of perfection and precise control"[33] are designing a simple *neuromorphic* "artificial electronic brain" consisting of a 2×2–millimeter mesh of silver nanowires, akin to a "highly interconnected plate of [nano] noodles"[34] (plate 6). It is only in its complex connectivity that the mesh of nanowires attempts to resemble the brain: it has 1 billion artificial synapses per square centimeter (still a few orders of magnitude less than a human brain). The electrical activity of this simple device displays *self-organized criticality*, a property unique to complex systems—a state in between order and chaos in which all parts are connected for maximum efficiency. This is a characteristic that has been observed in the human brain. The device is not programmed; rather, it is *trained to compute*, and its structure naturally changes and evolves as it performs the calculation. For example, scientists fed the device with the first half of a six-year set of car traffic data from Los Angeles, in the form of a series of pulses that corresponded to the numbers of car per hour. After hundreds of training runs, the output of the device predicted the statistical trend for the second half of the data set, which it had never seen. This kind of neuromorphic device aligns its own complexity to that of the phenomenon that it is trying to understand (or model), rather than trying to model phenomena with mathematical approximations. Like the brain, it does not separate processing and memory; the processing itself creates the memory in the structure of the mesh. These experiments show that complexly structured matter can compute. In other words, computing solutions to problems is a feat that ordinary matter can achieve if its structure becomes interconnected in particular ways, which we are starting to discover.

This kind of computing process also works at the cellular level (albeit in an immensely more sophisticated way) and other biological computing systems. A team of scientists based in Japan have shown that an amoeba-like, multinucleate plasmodium—the mobile stage of the slime mold *Physarum polycephalum*—can search and find solutions for complex computational problems[35] (plate 6).

When cells explore their environment and adapt to it through mechanisms such as those I described above in the context of mechanotransduction, a myriad of signals arriving from nanoscale proteins are collected constantly by the cell. The cell conducts and steers these signals through its complexly connected physical, chemical, and genetic networks and finally *computes* solutions resulting in cellular behaviors. Well beyond the capacities of the simple "nanowire spaghetti on a plate" device, evolution has optimized the computing powers of the cell over billions of years so that it can adapt and change the algorithm as it goes along.

Neuromorphic electronics still have a long way to go to become remotely as clever as a cell.

Roboticists are adopting a related strategy by constructing macroscale robots that mimic the neuron connectivity and sensing capabilities of small organisms. A good example is a demonstration by the collaborative OpenWorm project, which is currently attempting to create a virtual *C. elegans* worm *in silico* that uses machine learning based on the actual worm's neural network. OpenWorm attracted media attention in 2014 when the collaborators uploaded the bioinspired AI algorithm into a "neurobot" endowed with some sensing capabilities—and posted online videos of the very un-worm-like device behaving somewhat like a worm.

This kind of science may help us in the future to achieve increasingly sophisticated computation, technologies, and even AI, but it will also be useful in elucidating why our universe has created brains and minds so keen to understand space and structure, as the fascination of the film *Powers of Ten* eloquently testifies. The brain's hierarchical, complex organization, and its connections with the rest of the body all the way down to our cells, proteins, and DNA, make us both eager and able to explore the world around us. This kind of hierarchical assembly is the way evolution has consistently embedded in itself the capacity to learn and adapt, at the molecular, cellular, and organism levels. Bacteria, plants, animals, and humans explore, adapt, and learn because we are built by the rules that govern the universe. Some of the more audacious scientific theses even describe the whole universe as a huge cellular automaton able to evolve and compute its own behavior. Science has come a long way from the reductionism of the "central dogma."

BY EMBRACING BIOLOGY'S COMPLEXITY, SCIENCE IS CLOSING A HISTORICAL LOOP OF THOUSANDS OF YEARS

In learning the workings of biology and intelligence and uncovering their hierarchical understructure, modern science is reinvestigating something that humans have both practiced and intuited from ancient times: we are intensely interested in hierarchical structures and patterns—good at recognizing them, and also good at making and enjoying them. And we are not alone. Not only humans, but unicellular organisms, ants, bees, spiders, birds, monkeys, fish, and more, all build and understand networks and patterns. Humans are particularly hardwired to look for connections and patterns, to search for and identify pieces of the cosmic jigsaw puzzle, and to adjust the way we place ourselves in the universe by abstracting ideas from those patterns. Patterned structure and hierarchy are the underlying principles of aesthetics and art, music, language, science, philosophy, and even religious spirituality. From ancient times, humans have felt enjoyment, fulfillment, and reassurance constructing hierarchical, patterned explanations of reality. And we have also wondered where our interest in patterns comes from.

Early civilizations condensed their knowledge and intuitions about the universe into intricate cults and myths; Western and

Eastern religions and philosophical traditions held that the human ability to interpret reality emanated from God (often identified as the universe itself). Lacking science, they nevertheless intuited that living beings *emerged* mirroring the structure of the universe—or, as many ancient religions put it, we are created in the image of God, *imago dei*. The gods of patterns and hierarchical structuring live inside us, in our language and soul.

In the twenty-first century, modern quantitative biology, physics, computer science, neuromorphic electronics, and other branches of natural computing[36] and mathematics are in fact closing a historical loop in the human quest to understand life and nature—and getting us closer to consummating the primeval intellectual efforts of ancient civilizations. As we strive to discover the precise ways in which living, intelligent organisms contain the physics of the universe at the core of our biology, scientists and technologists (often unwittingly) reconnect with our ancestors to identify the plan of the universe itself inside human and nonhuman beings, though in a secular way. Another way of saying all this is that biology is entering the realm of physics. In its most fundamental form, physics is the quest to find the rules that construct and govern the universe in all its scales and manifestations, and to express them in mathematical language. Engineering then uses the knowledge that physics has encoded into mathematics to create technology. Biology has now become the natural ground where both physics and engineering are happening.

The physics perspective on biology is powerful not only because in the twenty-first century the knowledge hidden in biology's complexity will transform our technologies and health-restoring capabilities. Physics is powerful because it shifts the scientific culture from which we inquire into ourselves, our life, and our health. We are no longer the output of genetic computations; we emerge from the cosmos itself. This is why the modern technological revolution that extracts its energy and potential from life itself is bringing us back to the starting point of human thinking. Pure scientific inquiry is leading us to very old questions: What are we? How do we learn? What is our place in the universe? I will come back to these thoughts in the last chapter of the book and the epilogue, but before that we need to explore a bit more science.

In this chapter, I have summarized how the convergence of quantitative sciences is constructing a very different picture of biology, accepting and exploring its multiscale complexity, and abandoning the reductionism of the previous generation. In the next chapter, I move back to molecules to show how biology is utilized as both a material and a blueprint for construction in the new materials sciences of the twenty-first century. I will explore how the practical method of "learning (biology) by making (nanotechnology)" is being pursued. We'll begin with DNA and protein nanotechnologies.

2

LEARNING BY MAKING

DNA and Protein Nanotechnology

Biology's multitasking nanomaterials, proteins, are assembled with atomic precision in the cell during translation of the genome by the ribosome, a complex molecular machine that is itself made of protein and RNA. The ribosome adds amino acids in the correct order, one by one, to create a precisely arranged string that spontaneously[1] folds into a three-dimensional shape, one specific to each protein type in each living species—bacteria, animal, or plant. Folded proteins are then further assembled and transported by the cell machinery to create biological structures and functions. Nanotechnology is still far from controlling the assembly of molecules in such an efficient and elegant way, but for decades now, it has been learning how to use the natural self-assembly properties of molecules such as DNA and peptides to fabricate increasingly sophisticated nanostructures.

Self-assembly, the self-organization of nanoscale building blocks into patterns or structures without involvement of an assembler (human, cellular, or otherwise), occurs often in biological organisms. These processes often take advantage of complementarity of structures and/or forces between the molecules, which naturally orients and assembles the molecules in the optimal way using the temperature of the water environment as the source of energy to drive the process.

In this chapter, I review the emergence of nanotechnology made with DNA and proteins, its progress, and the applications in which it is already having an impact. DNA, RNA, and proteins have all been used for more than two decades as nano–building blocks to assemble designer architectures. DNA nanotechnology pursues the design and construction of any arbitrary shape using artificial DNA building blocks, and, more importantly, it attempts to give these structures nanoscale functionality. This functionality has been demonstrated in the performance of such complex roles as active template for the synthesis of complex molecules and programmable molecular computer. Nanostructures have even been tasked with regulating the functions of living organisms and targeting tumor cells.

The field of protein nanotechnology endeavors to achieve the same goals using artificial or natural proteins. This is a tougher task than doing it with DNA (proteins are combinations of twenty building blocks, the amino acids, while DNA is made from only four bases), but can lead to many more applications, because proteins can take on literally any three-dimensional shape and functionality at the nanoscale.

The last five years have substantiated an eagerly anticipated dream: the emergence of "protein designers," able to draft on their computers "post-evolutionary proteins" that don't exist in nature. The ability to use nature's tricks to redesign biology's building blocks opens up possibilities beyond evolution. Protein nanoengineering is poised to revolutionize many fields, from medicine to electronics and materials sciences, and to generate a deeper understanding of matter at the nanometer scale.

THE BIRTH OF DNA NANOTECHNOLOGY

From our twenty-first-century perspective, the idea of constructing things at the nanoscale using the complementarity of DNA's bases looks rather straightforward. But it took a great deal of previous knowledge, and a special point of view, just to realize that DNA *could* actually be used as a building block for complex shapes.

The history of these ideas can be traced back to the discovery of DNA's structure. First of all, it took more than twenty years to confirm that the right-handed, double-helical model proposed by Crick and Watson in 1953 was correct. It was in 1973 that Alex Rich verified the details of the double helix. Confirming the structure of DNA required the availability of high-quality samples of DNA molecules, as well as much-improved methods for crystallizing balky biomolecules and analyzing the data from X-ray diffractometers. Rich pursued his structural studies of nucleic acids, both DNA and RNA, for years, making groundbreaking discoveries that revolutionized the field and laid the foundations of modern biotechnology. A particularly important moment came in 1973, when Rich's lab succeeded in making the crystal and, using new computational methods, at last could resolve the structure in detail. After all that effort, there were no surprises. Watson and Crick were right. That is how science works: things must be proven. And often the effort to prove difficult things brings about unexpected dividends, because solving them requires thinking from different angles and making connections with other fields.

Apart from resolving their structures, making crystals of biomolecules for X-ray analysis brought a radical shift in the way scientists worked with and thought about the molecules. Starting in the 1920s, with James Sumner's early efforts to crystallize the enzyme urease, scientists struggled to make crystals out of molecules that had not evolved to be symmetrically patterned in space. This requires a special point of view; it requires seeing the molecule—its structure, its charge, its interactions with water and ions, temperature and pressure—not as a natural biomolecule working in a cell, but as a building block of a symmetrical, unnatural 3-D structure. As I discussed in the previous chapter, reductionist approaches to science often lead to creativity; once we discover the building blocks of something and how they work, we often find a practical use for them: technology ensues. The success of toys like LEGO is based on human love of building with blocks. This change of vision led the crystallographer Ned Seeman (one of the authors of the work that verified the structure of DNA in Rich's lab) to envision DNA as a building block of shapes that could not be found in nature.

After his successful work with Rich at MIT, Seeman went on to set up his lab at the State University of New York. By 1980 his potential for success and tenure was severely threatened by the lack of good single crystals. As he famously says now in his lectures, "No crystals, no crystallography . . . no crystallographer." Desperate at the lack of results, he turned to making computer models of DNA structures instead, and started to think of DNA as a building block.

Seeman was interested in a particular DNA that forms a cross, the *Holliday junction*.[2] Holliday junctions are a branched structure of DNA, comprising four double-stranded arms, that appears in nature during the replication of chromosomes in cell division.

Seeman realized in 1979 that these junctions could be made from synthetic DNA, and that if one managed to connect the junctions in the right way, it would be possible, in principle, to create artificial lattices and networks out of them.

Lattices and networks have been a human preoccupation from antiquity. In realizing that DNA could be used to make lattices, Seeman linked up with many millennia of human intellectual and artistic interest in the symmetry of space, and with ancient basket weavers' practical mathematical search for symmetrical structures useful for carrying things. The exploration of space and materials in art, science, and technology has always been intertwined. In the 1950s, in M. C. Escher's artistic explorations of the "language of matter, space and the universe," his recurrent motifs of tessellation and infinity were stimulated by crystallography.[3] One of the examples of his work on three-dimensional periodical patterning is *Depth*, a wood engraving that he made in 1955. *Depth* is an array of "robotic fish-aeroplanes" disappearing into infinite space; the depth effect is achieved, Escher himself wrote, by a "[r]hythmic positioning of each fish at the intersections of a cubic threefold-rotation-point system."[4]

Seeman was thinking of junctions made of four and six branches of DNA when he related his shapes to the fish in Escher's *Depth*, each with its head, tail, top fin, bottom fin, left fin, and right fin creating a six-arm shape. An acute analyzer of crystal structures could not miss the symmetries of Escher's robotic fish and its infinite copies in space. Could one make patterns with junctions of six branches using artificial DNA? Seeman recognized that it could be done if sticky ends to glue the structures together could be engineered into the structure. Such sticky ends exist: they were well-known in genetic engineering at the time, and they can be constructed from an overhang at the end of a DNA double

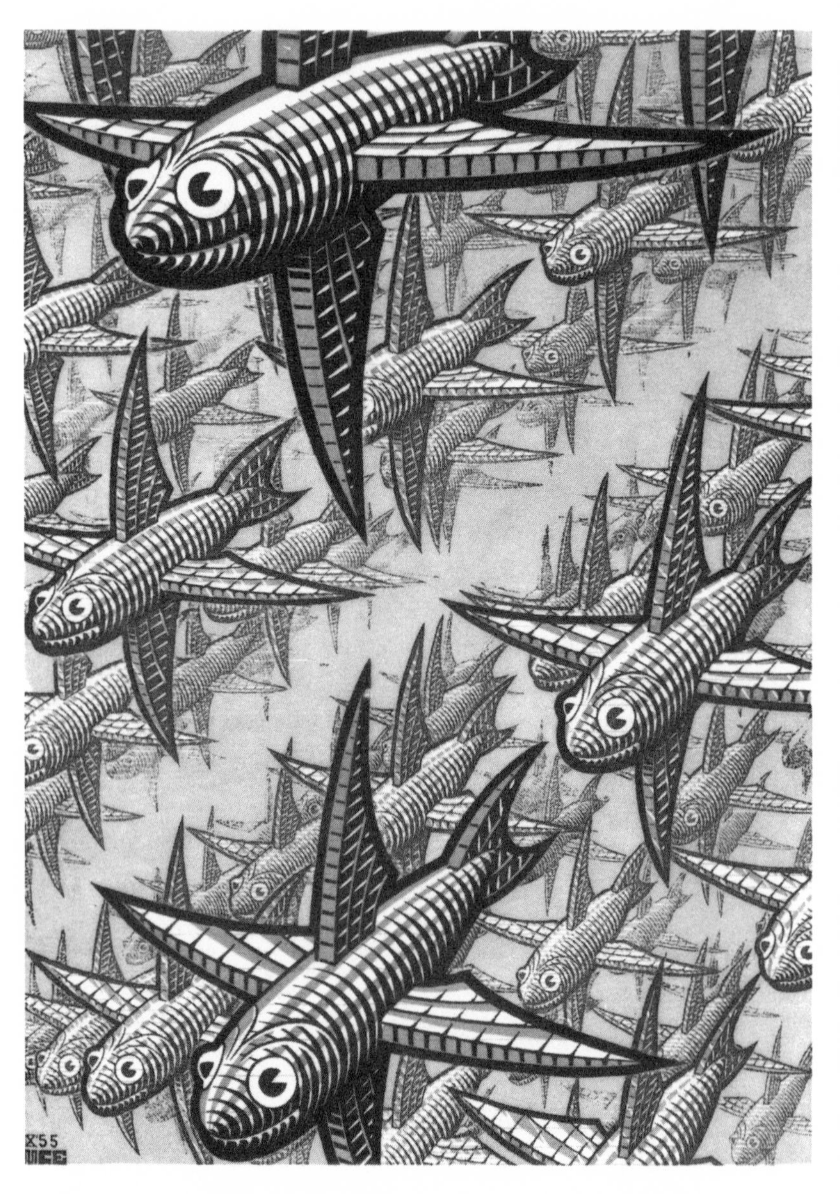

Figure 2.1. M. C. Escher's *Depth* (1955) was inspired by crystallography and later became the inspirer of DNA nanotechnology.

helix. Success came in 1988, when Seeman's lab used ten strands of DNA to create a molecule whose edges were connected like the edges of a cube. This marks the beginning of DNA nanotechnology. Seven years after the invention of the scanning tunneling microscope, three years after the discovery of carbon fullerenes ("buckyballs"), and two years after Eric Drexler's book *Engines of Creation*, the sciences were simultaneously making their way to the nanometer scale.

MAKING NANOSTRUCTURES WITH DNA

The DNA double helix forms naturally when two complementary DNA strands come into close proximity. This is because the four bases (also known as DNA nucleotides) that make up DNA (adenine, guanine, cytosine, and thymine) fall into two complementary pairs: A binds to T (and only T), and G to C (in biological notation, A≡T, G≡C). The complementarity of the DNA nucleotides confers on the DNA molecule the potential to become a programmable construction material, beyond being able to just make double helices. One could, in principle, construct any arbitrary shape by designing and synthesizing DNA strings in which the nucleotides are carefully positioned so that they bind to their complements in other strands, creating a nanoscale structure that folds into a particular shape. This is also possible because, apart from complementarity, a key feature of DNA is that it is structurally stable and robust for relatively long periods of time.

Ultimately, building with DNA strings would make it possible to construct any shape of potentially any size, with the structure designed down to the level of the DNA nucleotides. But to make this technology useful, it must be cost-effective as well.

For this reason, DNA nanotechnologists strive to find the best way of producing the shapes desired for specific applications. Eventually, the aim of the field is to produce both computer programs and DNA materials that anyone can use to create structures for their particular purposes. This will eventually facilitate the crafting of functional structures out of DNA much as architects and mechanical engineers now design structures in computer simulations, where they can test the suitability of a particular material for a desired task before it is manufactured in real life. DNA nanotechnology is currently developing both the building blocks and the software that will make this objective possible in the not-too-remote future. Always inspired by the holy grail of the ribosome, a main aim of the field is to create active structures, nano- and micro-DNA robots that can, for instance, synthesize molecules, or influence cell behavior in a therapeutic way.

The beginnings of the field were humble. It started with the construction of the cube that I mentioned in the previous section. The cube was followed by more-complex shapes, such as an octahedron. The initial research was an exploration of what shapes were possible, and it was marked by the determined effort of the pioneers to gain the knowledge necessary to advance. While the evolution of the field looks quite obvious in retrospect, it did not look so clear when the first PhD students were trying to build things out of DNA in the lab. One of the main hurdles was the vast amount of knowledge from different fields—chemistry, physics, microscopy, mathematics, and more—that needed to be brought together to succeed in making shapes.

After the cube and the octahedron, in the 1990s came knots made of DNA and RNA. With the knots, DNA connected with other ancient arts of constructing shapes using strings (e.g., crocheting and knitting), and also with the mathematicians' interest in topology.[5] The most sublime of knots, the Borromean ring, was tied with DNA in 1997.

All these structures were interesting but floppy, and this matters because you could not make lattices with them. If you want to construct large structures, one of the best strategies is to use tiles that are robust enough to self-assemble over relatively large areas. The next step toward constructing such tiles was to find rigid DNA motifs that could be used to build stable networks.

The breakthrough came with the creation of robust DNA double crossovers, or DXs, which allowed the first two-dimensional arrays of DX tiles to be constructed. Since then, two-dimensional arrays have been made from other motifs, including the Holliday junction rhombus lattice and various DX-based arrays. DX arrays have been made to fold up into 3-D structures, such as DNA nanotubes that can be connected to other structures. At the turn of the century, DNA patchwork and basketry using tiles was well on its way.

Scientists have always associated DNA with algorithms, and so the selectivity of DNA strings to preferentially bind to other DNA strings whose nucleotide sequence is complementary was immediately used to create two-dimensional DNA arrays able to implement computing programs. For example, DX tiles could be designed so their sticky ends would explore the possibilities of assembling different types of tiles into a lattice, so they act like Wang tiles (where edges of adjacent tiles must match),[6] allowing them to perform computation. A DX array encoding an XOR (Exclusive OR) operation has been constructed and demonstrated; its assembly allows the DNA array to implement a tessellation automaton that generates a fractal shape.

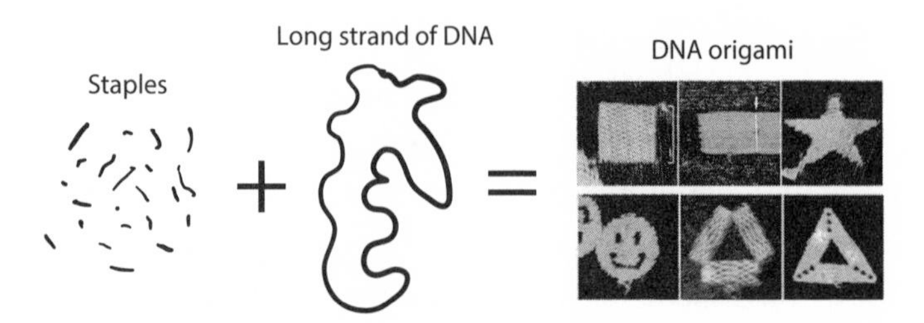

Figure 2.2. DNA origami. A long strand of DNA is folded by DNA staples that bind to specific locations, directing the spontaneous folding of the DNA strand into the desired shape, which is programmed in the design of the staples. Reprinted by permission from Springer Nature.

DNA ORIGAMI

A major breakthrough in the construction of planar structures with DNA was the realization that one could use small pieces of DNA as "staples" to stably fold long DNA strings. Paul Rothemund demonstrated in 2006 that any arbitrary shape could be made by folding a long DNA (usually a natural DNA strand from a virus) using computationally designed staple DNA strands, able to glue together bits of the long DNA strand so it can self-assemble into any arbitrary shape.[7] The staples are designed to be complementary to specific sections of the long DNA; as they bind they guide the folding process. The genius of "DNA origami" is additionally that it is easy and cheap to design and to build because the original DNA does not need to be synthesized. Rothemund's smiley faces made of DNA have become the unofficial logo of the field (fig. 2.2). Using an adaptation of this method, three-dimensional shapes have been created, too, such as DNA origami boxes that can be programmed to open the lid and release a cargo in response to a stimulus.[8]

DNA arrays have been used to add other molecules and create hybrid structures with proteins, carbon nanotubes, nanoparticles, quantum dots, and fullerenes to make molecular electronic devices. DNA arrays have been assembled on artificial lipid membranes, which offer their potential to create compartments and higher-order structures. It is also conceivable to attach DNA structures to membranes of living cells that can be used as handles on the cell to remotely modulate cellular function.[9]

DNA NANOROBOTS

The discovery of robust DNA structures that were not wobbly made it possible to fabricate DNA nanomechanical devices—DNA nanorobots capable of controlled movements. The idea of building a DNA nanomachine able to change from right-handed to left-handed DNA was conceived quite early in this field (1987), and it was in 1999 that the first DNA nanorobot was built. The first robots were nanowalkers, inspired by the walking behavior of protein motors, such as the myosin stepping motion I introduced in chapter 1. Several DNA walking systems have been demonstrated to date, using various strategies to generate and control their movement. Many of these DNA walkers use externally controlled DNA "fuel" strands (short strands of DNA that bind to specific places on the DNA robot, such as its "legs"), which then can be digested, for example, by enzymes. The energy produced in the enzymatic process is then used to generate movement of the DNA structure. DNA walkers locomote with strides a few nanometers in length, by advancing the trailing foot to the lead with each step. However, these early DNA nanorobots, because they were driven by molecular reactions, were very slow, had a low assembly or operation yield, or were unable to exert appreciable forces against external loading.

In 2010, Seeman's lab managed to make the first nanoscale assembly line. A DNA walker walking on a track embedded in a

DNA origami structure was able to pick up specific cassettes containing nanoparticles, move along the track, and assemble eight different products using three different cassettes.

In 2009, a DNA nanorobot made by DNA origami folding was built to release a drug payload when its aptamer (a piece of DNA that binds to a target molecule) sensors recognized specific small molecules.[10] The robot is a hexagonal prism box made of DNA that is latched closed with two DNA hinges; inside the box are twelve docking sites for therapeutic molecules. Two pieces of DNA that act as sensors open the box in response to cancer antigens. DNA nanotechnology has also been used to sense concentrations of chemicals in living cells[11] and to create and assemble chemical molecules that are difficult to make by traditional synthetic chemistry, as I will explain further in chapter 3 in the context of nanomedicine.

In 2018 a new way to control the robots has been successfully demonstrated: using electrical fields to move the different components of a DNA machine relative to each other. Doing this, the robot gains "many orders of magnitude in operation speed, almost perfect switching yield (precise, millisecond changes of position), and the capability of computer-controlled nanoscale motion and positioning."[12] The latest electric DNA robot is a 55 × 55–nanometer DNA-based molecular platform with an integrated robotic arm 25 nanometers in length, which can be extended to more than 400 nanometers. The arm can be used for electrically driven transport of molecules or nanoparticles over tens of nanometers. It is conceivable that in the future an array of these arms could be fabricated and remotely controlled to manufacture drugs—a nanoscale automated pharmaceutical plant.

SCALING UP DNA NANOTECHNOLOGY

DNA origami and tiles and cubes made of single-stranded DNA have proven to be the most useful technology for creating two- and three-dimensional objects by self-assembly. So far, however, the maximum size of the structures that can be achieved with these building blocks has been limited.[13] For example, one widely used scaffold for DNA origami is a genomic DNA approximately 7,200 nucleotides long, which folds into origami structures no more than 100 nanometers in diameter.

The other popular design strategy in DNA nanotechnology is the single-stranded tile assembly (SST), in which nanometer-scale 2-D rectangles or 3-D bricks made from single-stranded DNA are designed to interlock with each other. SSTs can be assembled into 2-D sheets or 3-D blocks that can be selectively "sculpted" to create different patterns and shapes, simply by including or omitting SSTs at specified locations in the structure. DNA shapes produced in this way are generally comparable in size to DNA origami nanostructures; larger structures have been built, but only at low yields.

In 2017, several research teams broke another barrier and managed to produce micrometer-size structures of DNA—orders of magnitude larger than those previously achieved—while at the same time scaling up the amounts that could be produced. Using a new computing algorithm, it is now possible to build up shapes using DNA origami square building blocks in a hierarchical, multistep process that assembles increasingly large arrays of the origami. The design software, FracTile Compiler, has been made available to the research community, so even non-experts can devise DNA sequences and experimental procedures to make large DNA patterns. The authors validated this automated design process by using it to make several "pictures" in DNA, including a *Mona Lisa* 700 nanometers wide.[14]

Another team has demonstrated that is possible to build millimeter-size structures using V-shaped DNA origami building blocks. Using sophisticated ways to control the assembly of the V-shapes, they have constructed three-dimensional tubes and several types of polyhedra.[15] Meanwhile, another group has succeeded in creating an SST DNA brick that can be used to assemble 3-D micrometer-size objects using their new software, "Nanobricks," as they demonstrated by creating various complex shapes, including a DNA teddy bear.

A lot of effort has also gone into reducing the cost of constructing with DNA. In 2017, advances in biotechnology have made it possible to reduce the price of DNA origami from $200 per milligram to around 20 cents.

All in all, the ability to construct any 2-D or 3-D shape out of DNA in an efficient and economical way is increasing at a very rapid rate. Any of the newest structures could be assembled into devices large enough to interact with biological cells—for therapeutic interventions, for example. It is conceivable that these structures can be used to program interactions between cells or other biological or inorganic objects, e.g. by designing scaffold materials that can connect such objects to create larger structures, such as an artificial biological tissue or an array of bacteria optimized to produce electricity. They will also enable the construction of molecular machines and assembly lines to synthesize molecules and polymers.

DNA nanotechnology has become an established field. Approximately sixty labs around the world are working on it, thinking and realizing shapes and functions, collaborating to find applications in medicine, computation, nanophotonics, and nanoelectronics. Open-source software is available that anyone can use to design a structure made of DNA; the DNA components needed to fabricate such structures are increasingly cheap; and the procedures to create many of the structures are relatively simple. These advances are accelerating the usefulness and the possibilities of DNA nanotechnology, which is poised to become a standard laboratory tool for atomic-precision construction in the not-so-distant future.

The main challenges to DNA nanotechnology applications in medicine are 1) degradation of DNA nanostructures when they are placed inside living organisms, and 2) the possible responses of immune systems to foreign DNA, since in many cases, such as DNA origami, the DNA is extracted directly from potential pathogens. Research is ongoing to improve the science that will enable the future utilization of DNA as a therapeutic agent.[16]

PROTEIN NANOTECHNOLOGY

The twenty amino acids that form proteins could, in principle, furnish a more multipurpose building material than the scantier four DNA bases. In fact, the polypeptide chains built of those twenty amino acids are the most versatile material on Earth. A folded protein is not only characterized by a stable shape; it is also carefully designed to have softness in the right place, the necessary interactions with water and other surrounding molecules and ions, the right assembly and disassembly properties to be capable of such feats as extracting energy from the environment to perform stepping movements, as myosin does. Beyond nano-walkers such as myosin and kinesin, proteins work as enzymes to catalyze chemical reactions; photonic nanodevices (e.g., rhodopsin in our eyes); pressure sensors (mechanosensitive channels); ultrasensitive chemodetectors of molecules (olfactory receptors); structural scaffolds (collagen in the extracellular matrix of tissues or tubulin and actin proteins forming key fibrillar structures of the cytoskeleton); nano–rotary motors (ATP synthase); protein factories (ribosome); electrical nanochannels and pumps . . . and more.

Natural proteins offer many possibilities for assembly, but their practical application to making arrays is only starting to be explored. In the previous section we saw what a challenge it has been for DNA nanotechnologists to construct lattices that expand beyond hundreds of nanometers. It has been shown, however, that it is relatively simple to construct regular two-dimensional lattices of clathrin protein that expand to several square millimeters (plate 7). Clathrin (from the Latin *clathratus*, or lattice-like) has three legs, forming a triskelion shape; in the cell it self-assembles into polyhedra that surround lipid vesicles, producing robust shuttles for transporting molecules across the cell. Clathrin's aptitude for assembly can be tailored to pattern a plane in large millimeter-wide sheets with a 30-nanometer periodicity, and it has been shown that these lattices can be useful for assembling enzymes and nanoparticles attached to the lattice.[17] Other natural proteins, such as ferritin, S-layers (proteinaceous lattices that coat the surfaces of some bacteria and archaea), and hydrophobins (which assemble at the surface of fungi), have also been used to create large 2-D lattices.

While making self-assembling lattices can be technologically very useful, the most fundamental aim for a nanotechnologist is to be able to utilize the versatility of the protein polypeptide chain to construct any arbitrary three-dimensional shape.

Is it possible to learn the mechanisms by which proteins fold into shapes and interact with other proteins, and then to use their design principles to engineer materials with atomic precision? To answer this question, I will succinctly review the efforts of the scientific community to understand how proteins fold into stable shapes inside the cell. Most proteins in an organism have a correct *native conformation* they must fold in; otherwise a disease associated with misfolding occurs, which can be fatal. (Alzheimer's and Parkinson's diseases are among such *proteopathies.*) "Disordered proteins" are exceptions to this picture, and actually use their disorder as their main functional feature, but I won't discuss them here. In the 1950s it was realized that the shape of a protein is determined by its sequence of amino acids and the environment. Under the right conditions of temperature, pH, and concentration of salts, the amino acid sequence folds spontaneously into a functional three-dimensional protein structure.[18] This is a very important finding, because it means that if one is clever enough it could be possible to predict the shape of a protein from its amino acid sequence, avoiding the costly process of determining the structure experimentally through X-ray crystallography, nuclear magnetic resonance (NMR) spectroscopy, or cryo-electron microscopy. The experimental determination of a protein's structure can take weeks or even months or years, and it is expensive, around $100,000 per protein. In contrast, protein sequences are discovered at a very high rate; there are more than 7,400,000 protein sequences available in the National Center for Biotechnology Information non-redundant protein database, but fewer than 52,000 proteins' 3-D structures have been solved and deposited in the Protein Data Bank (the international online repository of "resolved" protein structures).

But how difficult is it? The problem of simulating folding in the computer is that the unfolded polypeptide chain has an astronomical number of possible ways to fold: for a typical small protein, the number of possible conformations would be 3^{300} or 10^{143}. For example, a protein with 100 amino acids would have 99 peptide bonds; each bond has 3 stable configurations, and thus the chain could have 198 different bond angles. If each of these bond angles can be in one of three stable conformations, the protein can in principle fold into a maximum of 3^{198} different conformations (including any possible folding redundancy). If a protein achieved its final shape by sampling all the possibilities, it would require a time longer than the age of the universe, and perhaps it would take an algorithm even longer to compute the shape by exploring all the possibilities. But, in real life, proteins fold spontaneously in milliseconds, even microseconds if they are small enough.

Years of experimental research have proven that proteins fold into a sequence of intermediate states (the hierarchical construction strategy of nature discussed in chapter 1 appears to be functioning here too). At each step the protein folds into a relatively stable shape (i.e., a state characterized by an energetic minimum). This relative stability prevents it from moving and exploring all the possible conformations of the unfolded state.[19] This multi-step process from the sequence of the protein to the folded 3-D shape was immediately targeted by scientists looking for a computational solution to the protein folding problem. Over the years many strategies have been tried, with incremental success. But the target has remained elusive; probing all the conformations that a polypeptide chain could assume would take far too much time and computational power.

An important fact was soon recognized by the brave scientists trying to tackle this seemingly impossible task: since twenty amino acids in nature are available to make proteins, in principle one could build 20^{200} different proteins of a typical size. But the number of proteins that have been produced by organisms on earth is much smaller, on the order of 10^{12}. Ever since life on earth started from unicellular organisms, evolution has produced a much smaller number of proteins than are possible, and furthermore, the proteins that do exist are structurally related. Studying and comparing them, scientists have classified them into families. This allows us to predict, to some extent, the plausible shapes that a protein of unknown structure but known sequence can take, by comparison to other known structures in databases. This is not enough, however; the problem is particularly difficult in the case of proteins for which there is no template to compare to, because they have to be solved with *ab initio* ("from scratch" in physics slang) methods.

For years, the problem of protein folding was considered too difficult to be solved. But the protein-folding community has been very good at organizing to find ways to cut through this unmanageable problem. Trying to balance the overall goal of a field with the local competitive conditions for researchers' academic survival is not easy. In physics there is a strong history of collective organization, such as in the search for the Higgs Boson at CERN, or the sharing of telescope networks by the large international community of astrophysicists. But these fields are established around massive research infrastructure, so that collaboration and competition toward a common goal is facilitated by the place where science is being done.

The protein community found an ingenious way to spur simultaneous competition and collaboration: by holding two biannual challenges, the CASP (Critical Assessment of protein Structure Prediction) and CAPRI (Critical Assessment of PRediction of Interactions) competitions. Every two years the CASP organizers propose a challenge for the computational protein folding community: they present the sequences of a series of proteins for which structures have just been worked out experimentally by X-ray diffraction or NMR (nuclear magnetic resonance). The participants don't know the folded structure, but the CASP committee knows. CAPRI asks participants to predict the interactions that allow two proteins to stick together in a particular way. In the CASP eleventh protein folding competition in 2014, one submission from David Baker's team at the University of Washington on a large protein called T0806 came back nearly identical to the experimental structure. The head of CASP remembers that the evaluator reviewing the predicted structure sent an email straight off to him when he saw it: "Either someone solved the protein folding problem, or cheated."[20] This point marked the breakthrough that would open wide the possibilities not only for protein folding prediction, but for a real revolution in the way we understand matter. It is a before-and-after moment not only for biology, but for materials science, and hence nanotechnology.

David Baker's team did not cheat. In 1998 they had created Rosetta, a computer program that was already making better predictions than the rest, because it was very good at identifying short protein stretches for which a plausible structure was already known. Somewhat mimicking the natural protein folding process, these possible structures were identified and tested in an increasingly accurate way. Doing this requires massive computational power and so the team went on to create a crowdsourced version of the program called Rosetta@home, launched in 2005. Baker was following other scientists who thought that using the idle computational power of volunteers would be a good way to increase their capacity and also to engage with the public, as in the pioneering Galaxy Zoo, or Folding@home.[21] Scientists were increasingly realizing that amateur citizen scientists could contribute to scientific progress from home. What better way to do science in the twenty-first century than by engaging with the whole world?

There are currently more than 60,000 computers in the Rosetta@home network, processing at over 210 teraflops. Rosetta@ home was complemented with "Foldit," a computer program that allows the players to use a set of rules and their instinct to pin down the structures of proteins. When they are successful, the names of the Foldit players appear in the list of authors of the scientific papers. It is now estimated that there are about 1 million citizen scientists contributing to the resolution of protein structures, or the interactions of proteins with drugs or DNA. Apart from amateurs, more than 400 professional scientists are continually improving the Rosetta software, which is made available for free to the academic community and for a fee to companies. The revenue is put back into research. Success came from collaboration and the building of a community that interacts closely with the public. I would suggest that the lessons learned from this approach are probably even more important that the scientific breakthroughs that it brought about.

Baker and his collaborators were very quick to incorporate the latest developments in genetics[22] into their software, which led to the successful prediction of the structure of T0806. Until 2011, *ab initio* models had managed to predict just fifty-six proteins from the estimated 8,000 protein families for which there is no available template. Since then, Baker's team alone has added more than 900 proteins, and it is estimated that Rosetta's approach will work for 4,700 protein families. With the massively incremental amount of genomic data available, it is likely that any protein structure could be predicted in a couple of years.

In parallel to these efforts, the protein folding community has applied their algorithms and computational prowess to determining the way two proteins would bind to each other, paving the way for a revolution in the design of medicines that can bind to specific target molecules. The natural evolution of this computational epic is the implementation of artificial intelligence to help identify the pathways for folding success, learn from the process, and increase the speed of the predictions.

Advances are proceeding at breakneck speed in the expected way (especially from the point of view of our nanotechnology narrative). Once structures can be predicted, one can start to use the software in reverse: Can we envision a protein that does not exist in nature and then create it for a specific purpose? In other words, can we design with atomic precision a nanosize object that is able to fold into a previously determined shape; interact with water, ions, and other proteins; and even self-assemble with other proteins to make larger previously determined structures?

De novo design of "post-evolutionary proteins" is now the quest of the protein folders, who are evolving into protein designers.[23] The first results are astonishing.

Brian Kuhlman of the University of North Carolina, the codeveloper of Rosetta, built Top7,[24] the first globular protein to be designed with a fold not existing in nature, in 2003. Much of the effort in protein design has concentrated on fabricating ideal protein structure using helices, sheets and small loops (the basic structural motifs that can be found in natural folded proteins). The successful designs are then used to construct DNA molecules that code for them, reversing the process that occurs in biological systems. Then the DNA molecule that encodes the designed protein is inserted in a microorganism, such as *E. coli* or yeast, that fabricates the protein in large amounts in a process known as *recombinant protein expression*. Sometimes this does not work, and the proteins designed in the computer do not fold into the predicted shapes, or they aggregate when they are being expressed. In addition to Top7, several proteins have been produced in this way in the last four years. These artificial proteins were found to be extremely stable, and they had structures that were found to be almost identical to the original designs.

Other recent activities have concentrated on the design of proteins with internal symmetry in which a single idealized unit is repeated many times, resulting in structures that remain stable even at temperatures as high as 95° C. The set of equations that Crick developed in the 1950s predicting the existence of "coiled coil" structures in proteins have now been used to generate idealized bundles of parallel or antiparallel helices with different lengths, twists, phasing, and orientation. This approach has been particularly successful, and several designs—including a peptide (small amino acid chain or very short protein) that can bind to carbon nanotubes, parallel self-assembling helical channels, protein cages, and ion transporters—have been realized. Again, these idealized proteins are very robust, remaining folded in very high concentrations of chemicals and in temperatures up to 95° C.

The approach that is more interesting from a nanotechnology point of view is the design of specificity and complementarity of protein interactions—the quest to devise a protein that can assemble with another protein in the right orientation to build any desired shape. The success of DNA nanotechnology is due to the extreme specificity and complementarity of DNA bases: G binds to C and A binds to T. This specificity arises from the detailed molecular design of the DNA bases, each of which presents an array of atoms ready to form hydrogen bonds with the perfectly matching array on its complement. Baker's lab has demonstrated that it is possible to do this with proteins, too: they have designed proteins that can assemble into extensive networks held together by hydrogen bonds located with atomic precision. In the near future it will surely be possible to apply much the same "digital" approach to nanofabrication with proteins as with DNA, using modular hydrogen bond networks to code for specificity.

So far, the design of precise interfaces between protein subunits has permitted the fabrication of self-assembling cyclic structures, tetrahedra, octahedra, and open two-dimensional assemblies. Protein interface design has also been used to fabricate 3-D, one- or two-component assemblies with icosahedral symmetry and 60 and 120 subunits, respectively (plate 8).

Protein designers are already envisioning the potential of incorporating artificial amino acids for generating folds and functions that are not available in biological systems.

The main limitation of *de novo* protein design is that only a small percentage of the designed proteins can actually be built by microorganisms, due in most cases to insolubility of the protein or the formation of unintended folds and structures. But the close interaction of the computational effort with experiment will lead to the successful resolution of some of these problems, and probably along the way new ideas will arise for exploring the possible applications even further.

Several problems remain before these technologies can be made useful for large-scale manufacturing of materials, medical devices, and other applications, such as electronic components assembly. The two problems are: 1) scaling up of the production of the DNA and protein structures, and, crucially, 2) simple ways to interface the nanoscale with the macroscopic world.

Simultaneously with these advances, synthetic biology has been developing the use of microorganisms such as bacteria and yeast to manufacture molecules. Synthetic biologists would "all love to imagine a world where we could adapt biology to manufacture any product renewably, quickly and on demand," in the words of Northwestern University's Michael Jewett.[25] The combined efforts of synthetic biologists and protein designers will lead to an increased capacity to create proteins with all sorts of properties. Furthermore, manufacturing large DNA molecules with determined sequences is becoming cheap and easy. Synthesis machines can now churn out strings of several thousand base pairs, and it can be expected that soon we will reach the point where you can simply synthesize any DNA, regardless of length. This facilitates the process of fabricating the genes for protein manufacturing using microorganisms. The "radical abundance" predicted by Eric Drexler in his pioneering 1986 book *Engines of Creation* is getting closer to reality.

NANOSTRUCTURES THAT OPTIMIZE THEMSELVES THROUGH BIOLOGICAL EVOLUTION

The ability to design synthetic nanomaterials computationally and then optimize them through evolution might sound like science fiction, but it has already been realized.[26] Baker's lab has designed and fabricated artificial proteins that assemble into icosahedral shapes imitating the structures of real viruses. The synthetic capsids are able to package their own RNA genome. Using *E.coli* bacteria as host for this simplified version of viruses, the team has managed to get these artificial structures to evolve over several generations. Evolution has optimized these structures in terms of stability in blood and in vivo circulation time, giving them virus-like characteristics. This incredible advance opens a host of new possibilities for creating non-viral structures that could evolve the desired properties for drug delivery and other medical applications. Such synthetic nanodevices would avoid the safety risks, costs and bioengineering challenges associated with real viruses currently used in biomedical applications such as immunotherapies.

BUILDING BIOMIMETIC MATERIALS AND DEVICES WITH NANOTECHNOLOGY

If in the future we want to create, let's say, an extraordinarily robust and sustainable material that can grow and heal damage caused by the environment, we will probably need to rely on strategies that mimic the hierarchical organization of biological systems. The field of biomimetic materials is old,[27] but with the development of nanotechnology and advances in biological research, it is now growing rapidly. In the early days, discoveries were made by serendipitous observations of nature; the field is now evolving into a more systematic study of the material properties and design principles of natural tissues from the perspectives of materials science and engineering. Importantly, collaborations between the physical and biological sciences have made possible increasingly sophisticated multiscale "multiphysics" computer simulations and models that incorporate the key characteristics of the materials (e.g. structure, stiffness, chemical reactivity), which will inform and inspire future design and engineering.

Hierarchical organization is key to information transmission across the scales and to growth, healing, and adaptation in multicellular organisms. To survive and adapt they grow into shapes using nanoscale proteins, microscale cells and ultimately whole tissues and bodies. For instance bone consists of a scaffold made of collagen and other fibrous proteins woven into specific patterns where mineral calcium phosphate (hydroxylapatite) is deposited. By combining protein and mineral at the nanoscale, the bone achieves its particular mechanical properties, it is at the same time hard and flexible, able to resist damage and fracture produced by enemies or accidents, and prepared to grow and regenerate. In charge of the construction and remodeling of the bone structure bone cells (osteoblasts and osteoclasts), which collectively monitor the structure and health of bone tissue and respond to changes throughout life. There has been considerable recent research on the hierarchical organization not only of bone, but also of plants, seashells, spider silk, and the lotus effect (superhydrophobic surfaces) in leaves and feathers; on the optical micro- and nanostructures of butterflies' wings; on the exoskeleton of arthropods and the structure of sponges. While we are still very far from replicating the properties of living matter in artificial materials, the first biomimetic materials are already making their appearance. At Harvard, Jeffrey Karp's lab is making medical adhesives based on the tuning of van der Waals forces that geckos' micro- and nanostructured feet achieve (commercialized by Gecko Biomedical). Other researchers are making adhesives that mimic, for instance, the mechanical properties of spider silk. In 2018, a simple chemical and mechanical procedure that optimizes the arrangement of cellulose nanofibers has been shown to make it possible to convert any piece of wood into a high-performance material that has better mechanical properties than steel.[28]

A recent invention is Watermark Ink, or W-INK, a device that is able to identify liquids, and impurities in liquids, using a combination of strategies found in biology. The wings of butterflies owe their brilliant colors to their nanoscale structure. The surface of the wings contains networks of tiny pores, the size of which determines the color. Another animal, the brittle star, a relative of starfish, changes its color from black to white by tuning the position of pigmented cells inside light-focusing structures (akin to a biological lens) arranged in patterns on the star's back. Mixing both strategies, Joanna Aizenberg's lab at Harvard has developed a liquid decoder that responds optically to liquid infiltration into chemically modified arrays of pores. Depending on the liquid wetting the engineered surface, the structure of the pores will react to produce a different color, from which the liquid's composition can be inferred. The device fits in the palm of the hand, it can function without power, and it has been commercialized since 2013.

FUTURE DEVICES: QUANTUM PHYSICS MEETS BIOLOGY MEETS NANOTECHNOLOGY

Viruses have been explored for technological applications beyond medicine since the 1990s. The virus envelope, or capsid, is made of proteins assembled in a periodic array. In 2016 Angela Belcher's lab at MIT showed how to improve the efficiency of energy transport in solar energy conversion systems by connecting a network of chromophores (light-collecting molecules) on an ordered biological virus template.[29] By genetically engineering the viruses, they could change the spacing of the chromophores in the pattern, and they established a link between inter-chromophore distances and energy transport properties. Doing this, they were able to achieve quantum coherent energy transport at room temperature, and to tune its efficiency (fig. 2.32).[30]

This example shows how cutting-edge ideas about energy conversion in physics draw inspiration from the physics of plants' photosynthesis (which is thought to involve quantum phenomena[31]).Then, by engineering protein arrays, nanotechnologists are able to create patterns of chromophore molecules that come closer to the efficiency that plants achieve in nature. These are the kinds of "transmaterials" that will be fabricated in the future, interfac-

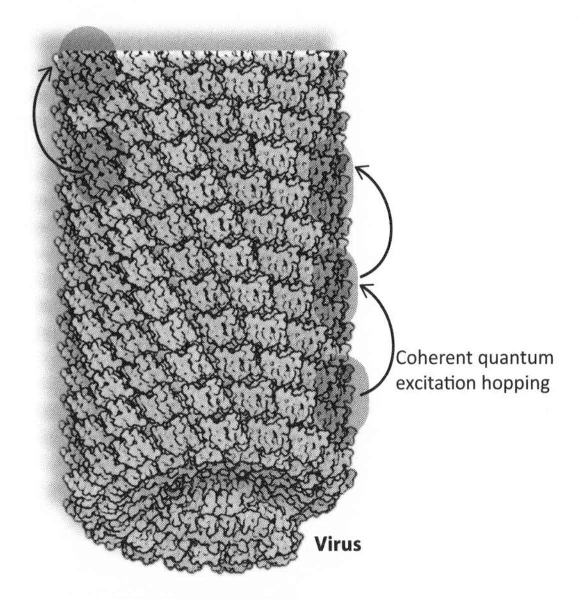

Figure 2.3. Virus genetically modified to achieve energy transport efficiency by tuning quantum coherence. After the virus is modified to adjust the spacing between the chromophores, energy can jump from one set of chromophores to the next faster and more efficiently.

ing advanced physics, biophysics, bio-nanotechnology, and protein and DNA nanotechnology for radically improved technological performance.

The next chapter continues the theme of "learning biology by making," but in a different context: medicine. In the introduction I explained how medicine acts as the integrator of all biological knowledge, and from some of the examples in chapters 1 and 2, it should be evident that medical application is very often the inspiration for new research. As new technologies appeared that allowed science to interact with matter at the nanometer scale, many moved to deploy them to solve medical challenges of our time, such as the development of new drugs and their targeted delivery to improve cancer chemotherapies, in a new field termed nanomedicine. The next chapter is my take on medicine's past and future from the precipice of a revolution in the understanding and control of our health brought about by the convergence of sciences in medicine.

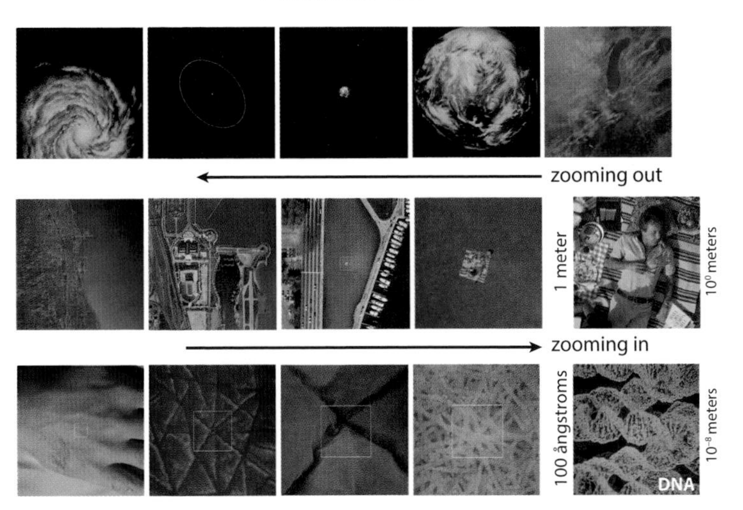

PLATE 1 The short film *Powers of Ten* (1968), by Charles and Ray Eames, shows the effect of adding a zero to the size of the field of view, multiplying or dividing dimensions by a factor of ten each time. *Powers of Ten* takes the viewer through the hierarchical arrangement of matter and forces in the universe, across the spatial scales from human size to the span of the whole universe, then back down the scales from galaxies into the skin of the human hand, to its DNA, and all the way down to the subatomic particles of the DNA's atoms. https://www.youtube.com/watch?v=0fKBhvDjuy0. © 2019 Eames Office, LLC (eamesoffice.com).

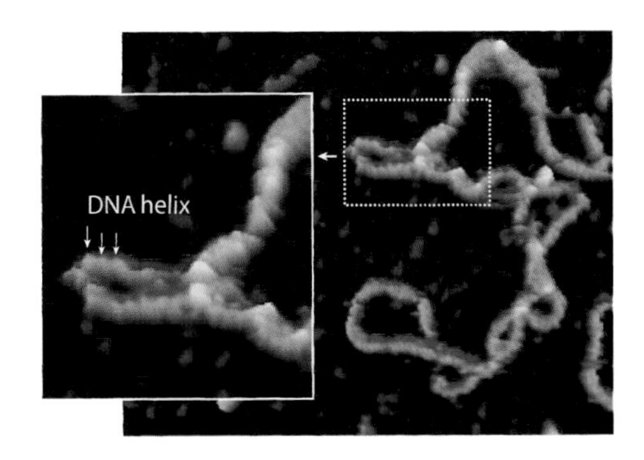

PLATE 2. AFM image of DNA extracted from an *E. coli* bacterium in a water solution. The inset shows the DNA double helix (three arrows). In living biological systems, DNA is twisted, a bit like the twisting of a circular rubber band; the tension imposed on the DNA stretches and compresses it locally. This tension can disrupt the double helix, as we see in the image. This kind of disruption possibly affects the expression of genes and opens a pathway for mechanical interactions with the world outside the cell. *Source*: Sonia Trigueros and Sonia Contera, 2008.

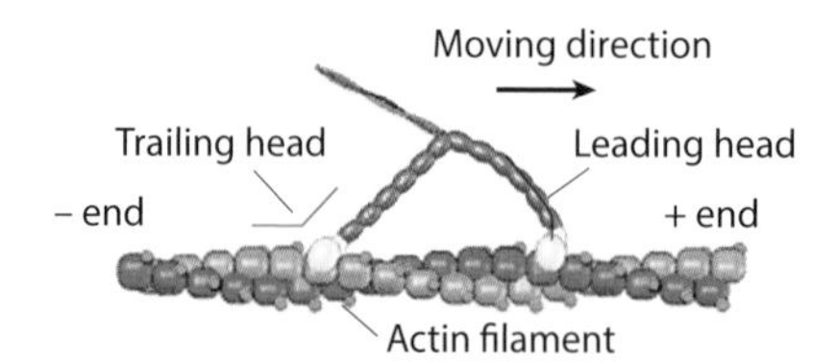

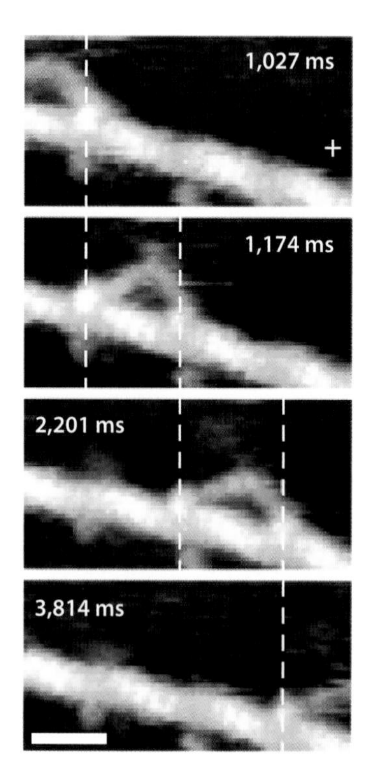

PLATE 3. High-speed AFM imaging of myosin walking on an actin filament track. Top panel shows a schematic representation of the myosin protein with its two heads stepping on the track. Successive AFM images show the processive movement of the protein. The scale bar is 30 nm. *Source:* Adapted from Noriyuki Kodera and Toshio Ando, "The Path to Visualization of Walking Myosin V by High-Speed Atomic Force Microscopy," *Biophysical Reviews* 6, no. 3–4 (2014): 237–60. AFM movies can be viewed here: http://www.nature.com/nature/journal/v468/n7320/extref/nature09450-s2.mov.

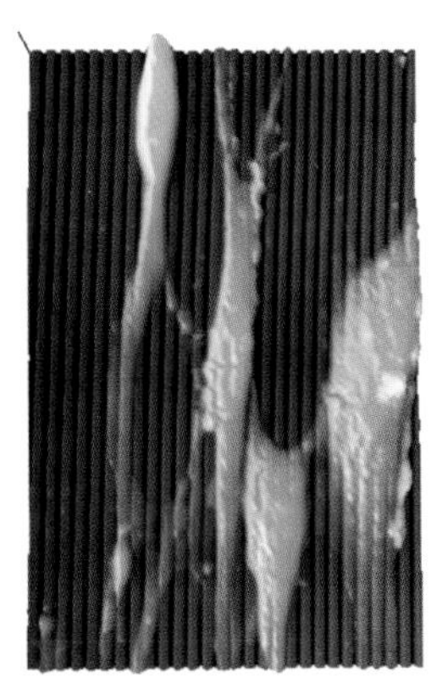

PLATE 4. Live mesenchymal stem cells as they attach and adapt to nanofabricated hard substrates with different nanostructures, imaged by atomic force microscope. The cells become elongated on the substrate with long grooves and more spread out on the substrate with nano-pillars; the cellular cytoskeleton aligns with the geometry of the nanostructures. *Source:* Sonia Contera, 2003.

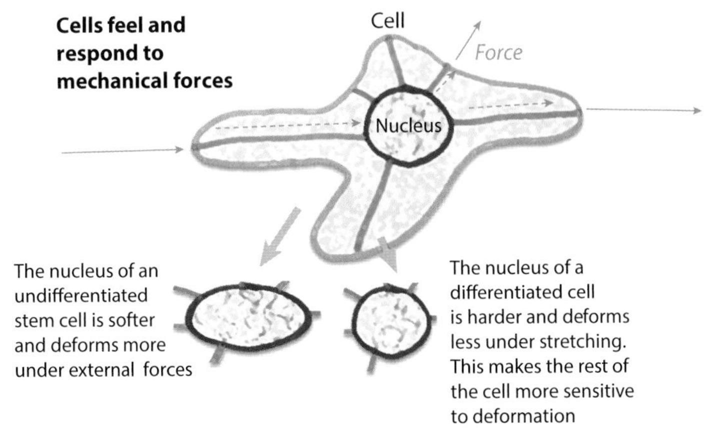

Cells feel and respond to mechanical forces
Cell
Force
Nucleus
The nucleus of an undifferentiated stem cell is softer and deforms more under external forces
The nucleus of a differentiated cell is harder and deforms less under stretching. This makes the rest of the cell more sensitive to deformation

PLATE 5. Cells feel and react to mechanical forces. The cartoon shows how the nucleus is connected to the external world by a complex network of nanoscale proteinaceous cables. The nucleus of a cell can be deformed by external forces acting on the cell. A stem cell nucleus is softer than the nucleus of a differentiated cell, and that has implications for the transmission of forces within the cell and to the DNA in the nucleus. In this way, it has been proposed that the cell nucleus acts as a *mechanostat* during stem cell differentiation.

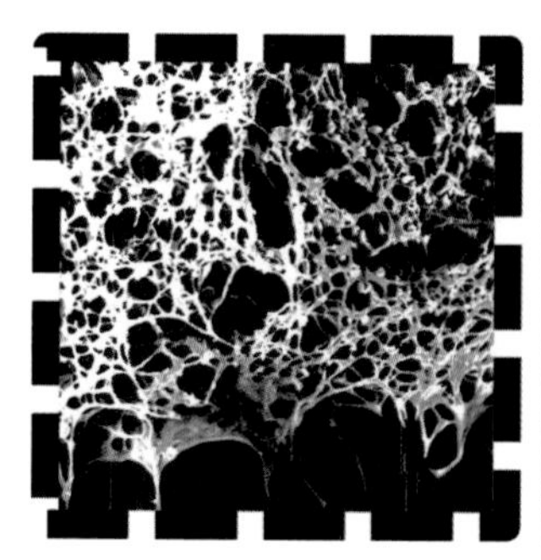

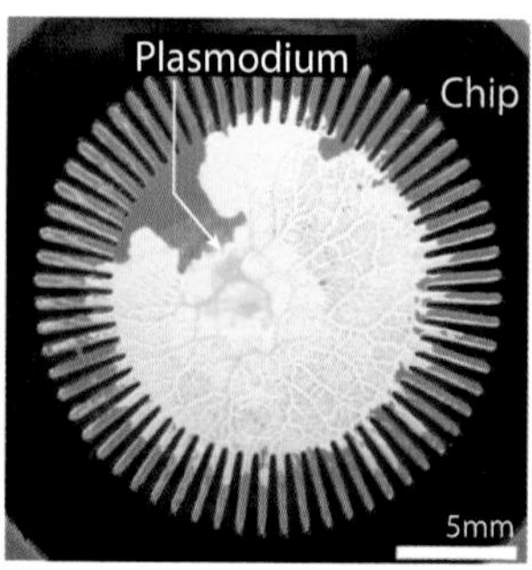

PLATE 6. Complexly interconnected matter (alive or not) can compute: *Left*, a mesh of highly connected nanowires can compute solutions to complex data about road traffic. *Source*: Sonia Contera. *Right*, the amoeba-like plasmodium of a slime mold calculates approximate solutions to a very hard math problem. *Source*: Masashi Aono.

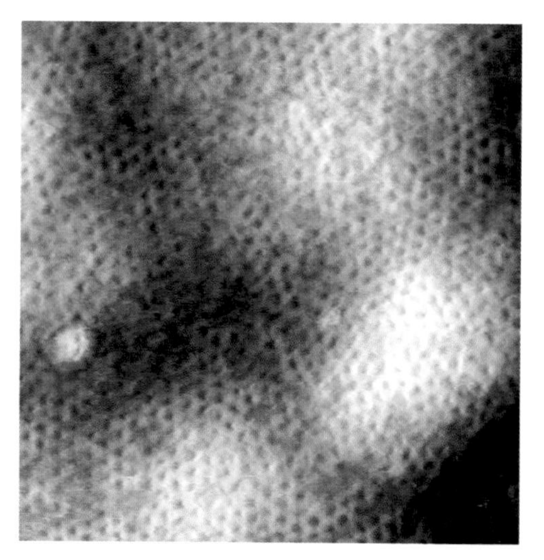

PLATE 7. A large hexagonal lattice assembled from clathrin proteins. The lattice was imaged in liquid with AFM. The image size is 990 × 990 nanometers. Courtesy of Philip Dannhauser and Iwan Schaap.

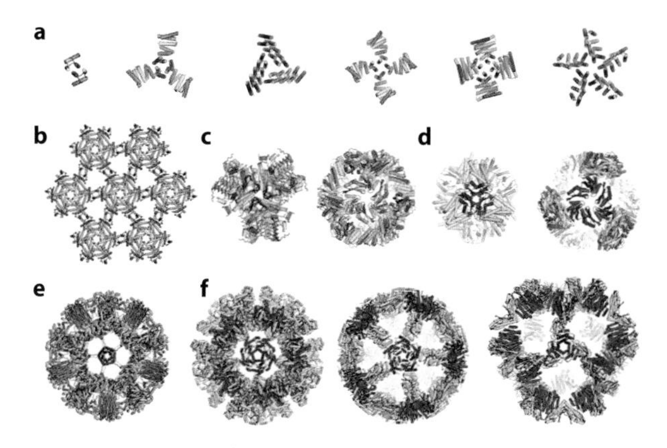

PLATE 8. Design of self-assembling proteins for the construction of self-assembling nanostructures: *a* shows the protein building blocks, and *b–d* show complex structures that can be built with them and other designer proteins. *Source:* Po-Ssu Huang, Scott E. Boyken, and David Baker, "The Coming of Age of De Novo Protein Design," *Nature* 537 (2016): 320–27.

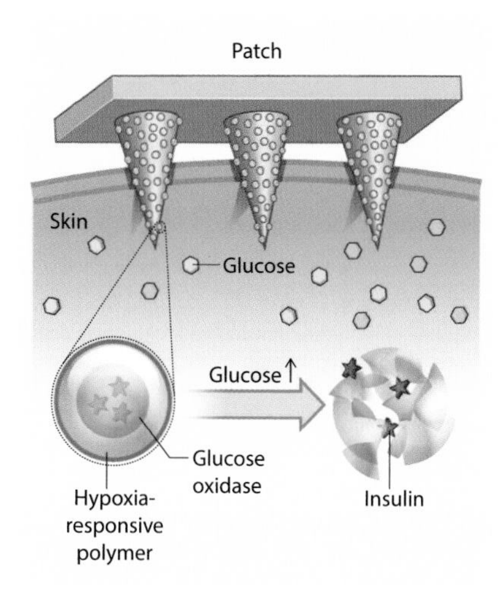

PLATE 9. A smart insulin-releasing patch made of nanoparticle-containing microneedles. The patch painlessly penetrates the interstitial fluid between subcutaneous skin cells. The nanoparticles in each needle contain insulin and the glucose-sensing enzyme glucose oxidase. Increases in glucose oxidase activity triggers disassembly of the nanoparticles and the release of insulin. *Source:* Omid Veiseh and Robert Langer, "Diabetes: A Smart Insulin Patch," *Nature* 524 (2015): 39–40.

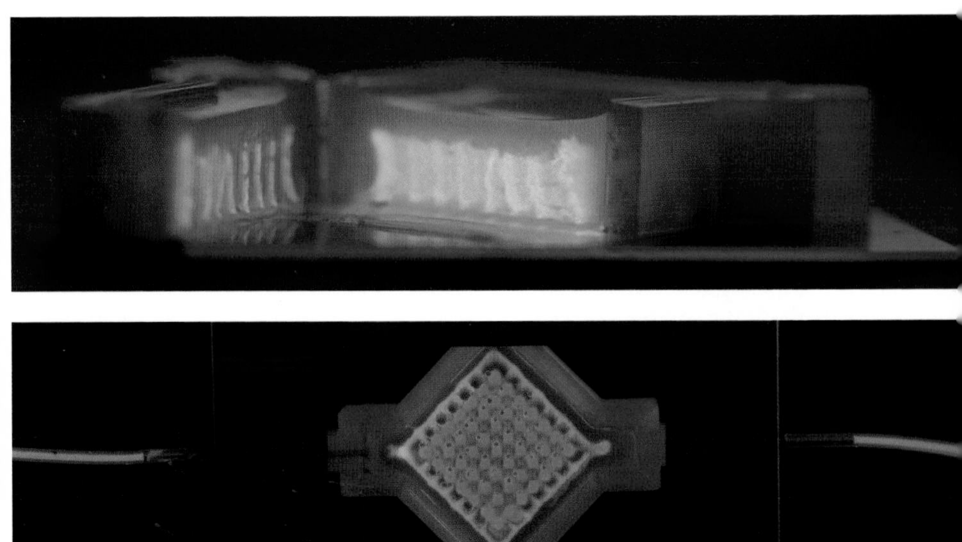

PLATE 10. Vascularized tissue constructs, 1 cm thick, made by advanced 3-D printing from human stem cells, extracellular matrix, and blood vessel endothelial cells. Top shows a picture of a longitudinal cut across the tissue. The vasculature within these tissues enables fluids, nutrients, and cell growth factors to flow uniformly throughout the tissue. Bottom shows the tissue in the organ-on-a-chip setup, which includes blood flow channels. *Source*: David B. Kolesky, Kimberly A. Homan, Mark A. Skylar-Scott, and Jennifer A. Lewis, Wyss Institute for Biologically Inspired Engineering at Harvard University.

PLATE 11. Robotic tissue-engineered stingray created by Kevin Kit Parker and collaborators in 2016. *Source*: Kevin Kit Parker.

PLATE 12. teamLab, *Enso*, 2017, Digital Work, Single channel, 18 min., 30 sec. (loop) © teamLab.

3

NANO IN MEDICINE

After decades of revolutionary drug discoveries—vaccination, antibiotics, statins, and other "miracle drugs"—that led to an unprecedented improvement in life expectancy and the conquest of many diseases, the turn of the twenty-first century saw a sharp decline in the number of effective medicines that pharmaceutical research was bringing to market. Science has responded to this concerning failure with increasingly sophisticated new multidisciplinary approaches for improving drug discovery, which are converging with the parallel advances of nanotechnologists. In this chapter, I will review the irruption of nanotechnology into drug design and delivery, and how it is expected to transform the effectiveness of treatments. As well as projections and predictions, I will survey early examples of how nanotechnology is actually beginning to be used to increase the success of leading-edge treatments such as cancer immunotherapies, gene editing, and gene delivery.

A BRIEF HISTORY OF DRUG DISCOVERY AND THE ARRIVAL OF NANOMEDICINE

Before the twentieth century, medicine did not cure very much. Over hundreds of years and generations, humans developed their traditional medicines along with all the other aspects of their cultures: their languages, their technologies, their writing systems, their visions and interpretations of the world. Through hit-and-miss experimentation and observation, they learned to identify herbs, plants, and foods that would alleviate certain conditions, and they passed on the knowledge they had acquired to future generations. Tea made from willow leaves (which contains "aspirin") was known to relieve pain, fever, and inflammation by ancient Sumerians, who recorded it on clay tablets four thousand years ago;[1] the use of sweet wormwood (in Chinese, *qīng hāo*) was known in Ming Dynasty China to be effective for treating malaria.[2] It took until the twentieth century to understand *why* many of these traditional remedies worked, and to derive effective, mass-producible treatments from the active chemical compounds in medicinal plants.

A crucial point in the development of modern medicine came when humanity recognized, accepted, and reacted to the fact that "germs" are the cause of many diseases. After the initial studies by Italian scientists in the seventeenth century (coinciding with the development of microscopes), it was not until the 1860s through 1880s that Louis Pasteur and Robert Koch developed a germ theory of disease that was widely acknowledged and trusted, as most people believed that diseases were caused by spontaneous generation. Pasteur had to be very creative to design a simple experiment that finally proved that "there is not a single known circumstance in which microscopic beings may be asserted to have entered the world without germs, without parents resembling them."[3] In 1867, Joseph Lister demonstrated that washing hands and cleaning wounds, surgical instruments, and operating rooms reduced deaths from infection in hospitals from 60 percent to 4 percent. Infection was not a chemical reaction of the exposed flesh in contact with air; it was caused by germs. From our twenty-first century perspective of rapid scientific advancement, it is incredible that medicine was in this state a mere 150 years ago; it is also a good measure of the speed of change. Cleanliness and disinfection made lifesaving operations worthwhile, where before, they had been ruled out due to the high risk of infection. It also led to reduced maternal and infant mortality around birth. Establishing the link between germs and disease also enabled Pasteur and others to speed up the development of vaccines—for cholera (1879), rabies (1882), tetanus and diphtheria (1890), and plague (1897).

Once germs had been disclosed and blamed for diseases, antibiotic substances started to be discovered, with initial reports appearing in 1877. Notably, Alexander Fleming discovered penicillin in 1929, as a result of observations on a mold that developed on some germ culture plates. But for penicillin to become widely available to everyone, someone had to determine the active compound in the mold, isolate it, demonstrate that it was not toxic to humans, and find a way to produce it in large amounts.

World War II acted as a strong catalyst for the development of modern antibiotics. In 1938, Ernst Boris Chain, a refugee from Nazi Germany, joined the Australian Howard Florey at Oxford to investigate natural antibacterials; their interest led them to revisit the work that Fleming had done with *Penicillium* fungi. As for the motivation of the work, Florey said, "People sometimes think that I and the others worked on penicillin because we were interested in suffering humanity. I don't think it ever crossed our minds about suffering humanity. This was an interesting scientific exercise, and because it was of some use in medicine is very gratifying, but this was not the reason that we started working on it."[4] Although Florey here contradicts my idea that a great proportion of biological research is inspired by medicine, he reaffirms something scientists love to repeat: blue-sky research is important, and leads to important transformations of the real world. This is a perspective that never stays in the sights of policy makers and research funders for very long.

In 1939, Florey and Chain, leading a team of British scientists and financed by a grant from the Rockefeller Foundation, identified the penicillin compound, started the first clinical trials, and led the successful small-scale manufacture of the drug from the liquid broth in which the mold grows. The timing helped to speed up production, as both the U.S. and British governments would provide assistance to ensure sufficient quantities of the drug to treat war wounds. During the war, the Oxford team got increasing economic and scientific support, and they managed to enroll one of the most talented crystallographers of all time, Dorothy Crowfoot Hodgkin, who was then a research fellow at Somerville College, Oxford. Hodgkin had prodigious skills for unraveling the chemical structures of complex molecules, in a time when computers could not be used to perform the laborious calculations. The larger the molecule, the tougher the calculations that had to be performed. The process had to be repeated many times; if the initial candidate structure proved not to lead to a diffraction pattern consistent with the experiment, then huge sets of calculations had to be repeated and refined until a good match was found.

On May 8, 1945, Victory in Europe Day, Dorothy Hodgkin made her way through the celebrating crowds in Oxford holding in her hands a delicate model of wires and corks. She had resolved the structure of penicillin and was trying to get to the Dunn School of Pathology to tell Chain. Hodgkin's structure of penicillin was, as Chain said in his 1945 Nobel Lecture, a great achievement not only for medicine but also for the determination of chemical structures from X-ray experiments: "[F]or the first time the structure of a whole molecule has been calculated from X-ray data, and it is the more remarkable that this should have been possible in the case of a substance having the complexity of the penicillin molecule."[5] Knowledge of the structure, together with parallel advances in synthetic organic chemistry, finally opened new avenues for creating and developing semisynthetic derivatives of penicillin—such as the cephalosporins—that sparked the creation of antibiotic treatments. Hodgkin herself was the recipient of the Nobel Prize in Chemistry in 1964.[6] As Florey explained, "Developing penicillin was a team effort, as these things tend to be." Success came from the combined techniques and points of view of scientists working in different disciplines with one application in mind. It is estimated that penicillin has saved over 82 million lives.

In parallel with biological research and X-ray crystallography, advances in synthetic organic chemistry played a fundamental role in the development of modern drugs. The first success was aspirin. It was known from antiquity that the bark and leaves of willow trees, genus *Salix*, can alleviate headaches, pain, and fever.[7] In 1853 the French chemist Charles Frédéric Gerhardt determined the chemical structure of salicylic acid and chemically synthesized acetylsalicylic acid. In 1897 the German chemist Felix Hoffmann, working for pharmaceutical company Bayer, possibly under the direction of Arthur Eichengrün,[8] found that adding an acetyl group to salicylic acid reduced its irritant properties, and Bayer patented the process. Aspirin was born. But it took until 1971 to understand why aspirin works so well. John Vane, professor of pharmacology at the University of London, described aspirin's mechanism of action as a dose-dependent inhibition of prostaglandin synthesis, which led to his sharing a Nobel Prize in 1982.

The rapid advances in chemical synthesis led to several successful decades for the pharmaceutical industries that had grown up around the new "miracle drugs," which were extending life expectancy for the privileged part of the world population that had access to them. In 1900, 1 in 3 deaths in the United States were caused by pneumonia, tuberculosis, and diarrhea. In 1940 the chance of dying from these diseases was 1 in 11, and by 2000, 1 in 25. The arrival of antibiotics, as well as drugs to combat hypertension and even cancer, has improved the health and life expectancy of millions.

Synthetic organic chemistry has been particularly successful at refining and optimizing the activity of known drugs which, in many cases, had their origin in biologically active natural products (as did aspirin). But this had to be done in a highly empirical, trial-and-error fashion. For most of the drugs that pharmaceutical companies developed in the twentieth century, molecules were just tried to see if they could be useful to cure diseases, without targeting any specific protein or receptor.

Beginning in the late nineteenth century, a series of key insights and discoveries allowed scientists to consider the possibility of strategies for drug design more targeted than just looking for fortuitous findings in nature, or in the chronicles of ancient traditional medicines. One such major advance was made by Paul Ehrlich in the 1870s. Ehrlich, a student and a friend of Koch, was the first to realize that there were macromolecules in the cell that acted as "chemoreceptors." Investigating the effect of coal-tar derivatives, particularly dyes, on living cells, he argued that certain chemoreceptors on parasites, microorganisms, and cancer cells were different from those in host tissues, and that therefore, in principle, they could be targeted with "magic bullets." In other words, Ehrlich invented chemotherapy. According to the database www.drugbank.ca, state-of-the-art pharmacopeia currently addresses about 60 proteins, of which 70 percent are proteins associated with the cell membrane and 28 percent are enzymes—i.e., most of the molecular targets of drugs are located at the cell surface.

The example of aspirin shows that finding out why a drug works is not easy. It requires a thorough biological knowledge of the drug's interactions with proteins and other molecules in the body—knowledge that is slow and laborious to obtain. Designing one drug from scratch to target a specific disease is even more complicated. For this reason, even now most drug development is done by trial and error. A lot of the biochemical, biological, and molecular cell biology research of the twentieth and twenty-first centuries has dealt with the molecular origins of disease and the identification of molecular targets to kill pathogens or misbehaving cells. Even when targets are known, the development of drugs that effectively reach and bind to a target is also very complicated. Detailed information about atomic structures is increasing exponentially, as is the power of computer simulation and analysis, but "rational drug design and delivery" is still out of reach, mainly because of the complexity of biological structures and the multidisciplinary nature of the problem. The pharmaceutical industry largely abandoned the "rational design" strategy in the 1990s to develop new empirical methods based on small molecule library synthesis and high-throughput screening of drugs in cell cultures. Though these methods use sophisticated robotics and biological techniques to assay thousands of compounds simultaneously, they add up to . . . more trial and error.

But finding drugs by "brute force"—testing all the compounds in existing libraries—has also failed to deliver results, perhaps as a result of a lack of chemical diversity in the large industrial libraries used for screening. New ideas for opening the pipeline are needed. An integration of the new quantitative biology explored in chapter 1 with mathematical models, computational techniques, engineering, and nanotechnology is expected to lead to the next stage of drug discovery, but the field has gotten stuck. Pharmaceutical companies are big and have enormous inertia; an entire community of science has been built around this slow and costly paradigm. Change is difficult when institutions and activities are very big. In the twenty-first century, however, as we look deeper into the biology of diseases and, importantly, gain the ability to model them quantitatively, it becomes much more possible to base new ideas for therapies on biology, physics, and rational design. And some of the most interesting advances are happening in the multidisciplinary arena that nanotechnology helped to create. Techniques and concepts from the physical sciences are contributing to the creation of new paradigms for fighting disease that encompass not only chemistry and biomolecular structures, but new methods borrowed from physics, engineering, and mathematics.

ANTIBIOTIC RESISTANCE AND NANOTECHNOLOGY

The push we need to start developing a new, more rational and agile pharmaceutical industry in the twenty-first century may come, once again, from bacteria—from their ability to become resistant to antibiotics. Most of our antibiotics are derived in one way or another from the molecular weapons that fungi and plants have developed over millions of years of warfare with bacteria in their natural ecosystems. Over those millions of years of constant, rapid evolution, bacteria have produced an amazing array of weapons and strategies for neutralizing the threat of antibiotics. Bacteria can produce enzymes that inactivate an antibiotic drug; they can change their surface or their cell wall to inhibit the drug's uptake; they can alter the protein, enzyme, or receptor targeted by the drug; and they can even activate drug efflux pumps that deliberately extract and remove the drug from the cell, acting like molecular vacuum cleaners just for antibiotics.

Biological scientists have discovered that bacteria evolve resistance by using two types of genetic mechanisms. The first one is mutation. Mutations in the DNA of the bacterial genome occur naturally when cells divide. Bacteria are especially prone to mutation, because their genome consists of a single chromosome and because they can divide very rapidly. Through mutations, a bacterium can "discover" and express a tweaked protein that foils the action of the antibiotic, survive, and then pass the tweak on to its daughters. Bacteria can also take in or transmit small pieces of circular DNA called plasmids that may carry mutations. One possibility is that a bacteriophage virus inserts a plasmid from a bacterium it previously infected into its next victim, and so genetic material passes from one to the next by *transduction*. But perhaps the most powerful and surprising weapon is *conjugation*. During conjugation, two bacteria that come into physical contact with each other can transfer a plasmid carrying genes for enzymes that inactivate antibiotics. In this way, the information needed to become antibiotic resistant can be passed very rapidly from one bacterium to others. To make things worse, plasmids carrying antibiotic-resistance genes can be transferred *between different species* of bacteria, making the wide spread of resistance to a specific antibiotic through a bacterial population very rapid and efficient. Bacteria have been evolving on earth over billions of years, and they have probably the most efficient, quick, and clever solutions for evolving out of almost any danger and adapting to any environment.

Conjugation that spread a plasmid carrying the gene for resistance to methicillin (an antibiotic derived from penicillin) is the prime suspect in the origin of the best-known superbug, MRSA (methicillin resistant *Staphylococcus aureus*). Penicillin and methicillin work by weakening the bacterial cell wall, which eventually results in the bacterium breaking open, or *lysing*. In MRSA, the plasmid transmitted through conjugation encodes a protein capable of inhibiting methicillin binding. Another example is a plasmid carrying a gene for the enzyme beta-lactamase. Beta-lactamase alters the structure of the penicillin molecule, rendering it inactive. Use and abuse of antibiotics since the 1950s, to treat human diseases and also to control infection in factory-farmed livestock, has created the potential for a public-health catastrophe: a "post-antibiotic" era in which no antibiotic drug is effective against infection. Pharmaceutical companies have not considered antibiotics economically worth the investment in research and development that has to be justified by shareholder value. Confronted with this alarming situation, governments worldwide are issuing calls and funding initiatives to find new paradigms and drugs that will prevent a return to the death-by-infection statistics of the nineteenth century. Alongside the old strategies—identifying more natural compounds that have antibiotic activity; modifying existing drugs to make them newly potent; trying new combinations of existing drugs—the new multidisciplinary teams are focused on finding new paradigms, new targets that make it more difficult for bacteria to develop resistance.

For instance, the quintessential tool of nanotechnology, the AFM, is being combined with superresolution fluorescence microscopy to study the structure and assembly of the bacterial cell wall, the *sacculus*, which is made of sugar-and-amino-acid molecules called *peptidoglycans*. The sacculus is a dynamic and adaptable structure fundamental for responding to the environment and for signaling, and thus a useful target for antibiotics. Until 2013, the architecture of the sacculus was not known.[9] The peptidoglycan mesh, made up of circumferentially oriented bands of material interspersed with a more porous network, has produced preliminary hypotheses to explain bacterial growth that may conduce to the discovery of new antibiotic targets. Laboratories including my own lab have developed methods for measuring the mechanical and even electrical properties of bacteria. Acquiring physical information about bacteria opens up new possibilities for the design of antibiotics. Rather than just targeting molecules and biochemical pathways, with the combined knowledge of biology and physics we will be able to develop strategies such as targeting architectural patterns or the dynamics or assembly of the sacculus, and disrupting the stiffness and electrostatics of the bacterial surface. Bacteria swim, stick, communicate, and feed themselves using physical means as well as chemical, and the new knowledge generated by nanotechnology and quantitative biology will allow us to move on from chemical battles toward more comprehensive physico-chemical warfare against resistant bugs.

The first successful strategy to identify and use a physical target for antibiotic design has been achieved by the group of Gerard Wong at UCLA, targeting bacterial *persisters*. Persisters have reduced metabolic activity and are tolerant to antibiotics that target growth processes, such as cell wall, protein, and nucleic acid synthesis. Persistent bacteria can revert to an actively growing state after antibiotic treatment is ceased, and in this way, they are a major factor in recurrent and chronic infections, as well as a reservoir of resistant mutants that can increase antibiotic resistance. Wong's group has succeeded in engineering a prototypical persister-specific antibiotic by adding persister-killing properties to an existing antibiotic, tobramycin. Tobramycin is a potent antibiotic, but had limited activity against persisters. The Wong lab decided to redesign it using new knowledge about the shape and the topology of the bacterial cell.

In the last five years, several structural studies and simulations by Wong's lab have demonstrated that a nanoscopic negative Gaussian membrane curvature (a way of measuring curvature in mathematics) is topologically required for membrane permeabilization, such as the formation of pores, in bacteria. The Wong lab had shown that the ability to generate this nanoscopic membrane curvature is "programmed" into some peptide sequences by the way that patterns in these peptides' physical properties relate to the geometric requirements of membrane curvature changes. Armed with this model, they designed "Pentobra" by adding a 12-amino-acid negative Gaussian curvature–forming sequence to tobramycin, effectively creating a multifunctional antibiotic that combines the membrane-penetrating activity of the peptide with the drug's inhibition of protein synthesis. In 2014 they reported that Pentobra permeates membranes and retains high antibiotic activity. The two mechanisms work with one another very strongly, and that may be what makes this antibiotic so powerful: Pentobra can kill persister-cell strains of *E. coli* and *S. aureus* 10,000 to 1 million times better than tobramycin, but remains nontoxic to eukaryotes. These results suggest one promising strategy: to renovate traditional antibiotics by combining several strategies in one molecule. In 2015 they demonstrated the successful use of Pentobra in the treatment of bacterial acne.

The website of Wong's lab summarizes not only the philosophy of his research, but the essence of the new biomedical paradigm emerging in the twenty-first century—the convergence of disciplines: "[W]e use a broad range of tools, including synchrotron X-ray scattering and spectroscopy, quantitative optical microscopy, microfluidic systems, machine learning and 'big data' methods, X-ray and electron microscopy, and laser scanning confocal microscopy . . . [O]ur collaborations include not just bioengineers, but also medical doctors, biologists, physicists, chemists, and material scientists."

Another strategy is to look for radically new ways to prevent antibiotic resistance from evolving in the first place. This often involves the design of nanoparticles with special characteristics that make them lethal and, at the same time, selective of harmful bacteria. Bacteria are much more negatively charged than most cells, and so positively charged polymers accumulate at their surface. Research has shown that nanoparticles that self-assemble in vivo from polycarbonate polymers can penetrate and disintegrate the bacterial cell wall and membrane, killing the bacterium.[10] The polycarbonate material is biodegradable: it can be broken down by enzymes in the body, and thus is less likely to accumulate. After a short period of time, the nanoparticles and their constituent polymers should revert back to an innocuous by-product with a very low molecular weight, which can easily be cleared.[11] These polymers are cheap and easy to produce, and have significant potential as future infection fighters.

Long before penicillin became the antibiotic of choice, silver was used as an antibacterial agent. Hippocrates was discussing the use of silver in the care of wounds 2,400 years ago. In the early twentieth century, silver was used in eye drops and sutures, and during World War I soldiers used silver leaf to treat infected wounds. Although modern research has shown that in most such cases silver is an inefficient antibiotic agent, its earlier use has inspired research into the antibiotic properties of silver nanoparticles. A large body of research in the twenty-first century has shown that silver nanoparticles indeed present wide-ranging antimicrobial, antifungal, and antiviral activity; they have proven effective at killing bacteria such as *E. coli* and *Pseudomonas aeruginosa*, including antibiotic-resistant strains. The exact mechanism by which they kill bacteria is the subject of much debate: Is it their shape, their size, or the silver ions they leak that destroys the bacteria, or a combination of them all? Many applications of antibacterial nanoparticles are being investigated, and they are already common in consumer products such as socks, carpets, and water bowls for dogs.[12] However, there are concerns about their toxicity to humans, animals, and the environment, which are currently being addressed in extensive research programs in Europe, the United States, and Asia.

An alternative, environmentally friendly approach using nanoparticles of lignin, a natural polymer present in the wood and bark of plants, was reported in 2015.[13] The lignin nanoparticles were infused with silver ions and coated with a polymer that becomes positively charged on contact with water. This work shows that it is possible to create a nanoparticle that combines high antimicrobial activity with low environmental impact using the principles of "green chemistry" (a new branch of chemistry that develops technologies to prevent pollution and reduce consumption of nonrenewable resources).

Another interesting aspect of using nanoparticles to develop antibiotics is that bacteria have not evolved in an environment with artificial nanoparticles, so it would presumably be more difficult for them to evolve resistance. Recent research has demonstrated that silver nanoparticles coated with polymers can have a strong antimicrobial action that bacteria cannot evolve resistance to. These particles also seem to promote wound healing.[14]

Many hurdles need to be overcome in the development of a totally new kind of antibiotics that prevent the advance of bacterial resistance—the antibiotic research challenge of the twenty-first century. The scientific community is now using the tools of biology, nanotechnology, physics, chemistry, engineering, mathematics, and computer science to craft new strategies and find new paradigms in increasingly multidisciplinary and international teams. Another significant innovation in this effort is the "responsible" way science is being done in the labs to avoid unintended consequences, such as environmental damage.

RATIONAL DRUG DESIGN USING DESIGNER PROTEINS

Modern efforts toward rational design of medicines are not limited to antibiotics. The new protein designers introduced in the previous chapter have a very strong motivation to use their newly acquired powers to design new drugs. Computational methods based on the peer-to-peer computing platform Rosetta@home were used in 2011 by David Baker and colleagues at the University of Washington to design new antiviral proteins not found in nature. These artificial proteins target specific molecules on the surface of flu viruses, with the purpose of blocking the virus's invasion of the cell and its reproduction. Their approach can be seen as akin to engineering a small space shuttle that is able to break the secret code, dock with the enemy's space station, and destroy it—except that to do this with a virus, it is necessary to engineer the docking mechanism and code at the atomic scale. It took almost eighty rounds of computing efforts to design a protein that targeted multiple strains of flu, which could then be produced by yeast cell cultures.[15] Eventually the team succeeded, and in 2016 they proved that their molecule was ten times more effective than Tamiflu.[16] The designer protein binds the virus surface molecule so tightly that the lab dubbed it "Flu-Glue." It is small and stable, so that it can be formulated for delivery to the lungs by aerosols through the nose. Delivered repeatedly to mice's lungs, this anti-flu protein has proven safe, and able to protect animals from lethal doses of flu virus. The initial results in mice seem to indicate that it does not provoke any significant immune response. In a common move of modern biomedical research, once an idea shows potential, someone will try to turn it into an industrial activity. Aaron Chevalier, a former Baker lab Ph.D. student, is now one of the leaders of Virvio, a Seattle biotech company that is working to commercialize the protein as a universal anti-flu drug.

Besides killing viruses, Baker's team at the Institute for Protein Design (launched at the University of Washington in 2012) has discovered that their designer protein can *detect* viruses. They insert "Flu-Glue" into low-cost nitrocellulose paper membranes to make viral infection diagnostic test strips that are very sensitive. Their paper strips are able to detect fewer than 100 flu virus particles in a single nasal swab.

The ability to design proteins to target problems is inspiring the creativity of students. In 2011, a UW student team won the grand championship prize at the annual International Genetically Engineered Machine (iGEM) competition. Led by Ingrid Swanson

Pultz, the students used the computational design powers of Baker's lab to reengineer the specificity of an enzyme so that it would zero in on and break down gluten in the harsh acidic conditions of the stomach. They called their enzyme KumaMax after the enzyme they modified, kumamolisin. In 2015, they produced an advanced version of KumaMax, Kuma030, which breaks gluten proteins at the precise places that are known to cause an improper immune reaction in the small intestine of people who suffer from celiac disease. Kuma030 has high enough activity in the stomach to ensure that a small tablet taken before a meal can completely break down any accidentally ingested gluten. Kuma030 is now being developed by a spinout company.

The power to design and build proteins has also been applied to the development of vaccines. In 2010, Baker's lab proved, using Rosetta, that proteins designed to approximate parts of the AIDS/HIV virus can promote the production of neutralizing antibodies to HIV when they are injected into animals as vaccines. Using the same strategy in 2014, they created a new protein for immunization against the respiratory syncytial virus (RSV), which is particularly deadly to infants and the elderly.[17]

Most of these new protein designs are currently tested in animals to assess their efficacy, their side effects, and their possible toxicity. Developing a drug up to the point where it is suitable for use in humans takes many years of preclinical and clinical trials; many strategies have to be abandoned or redesigned. But the combination of techniques and ideas from all the available sources of knowledge can be expected to bring major breakthroughs in the coming decades.

DNA NANOROBOTS FOR PROGRAMMABLE CHEMICAL SYNTHESIS

The DNA nanotechnologists introduced in the previous chapter have been working since 2002[18] to create molecular assemblers that can be programmed to construct any desired large molecule, including proteins. This is highly relevant to drug discovery because one of the bottlenecks to producing new drugs is insufficient variety of the molecules in chemical libraries, and expanding those libraries requires new synthesis methods. Chemical synthesis is usually done by a series of chemical reactions between several compounds. Achieving this for the construction of large molecules in sufficient amounts is very difficult. Indeed, biology needs a molecular assembler, the ribosome, to construct the proteins in all living organisms. In recent years, DNA nanorobots inspired by the ribosome have been developed—molecular-scale assemblers that in the future will be programmable to construct any desired molecule. The ribosomes are able to build all the proteins in an organism by selecting building blocks from a pool of RNA-linked amino acids and concatenating them in a sequence defined by the messenger RNA molecular program. Mimicking the "block by block" assembly method of the ribosome, DNA nanotechnologists are getting much better at constructing complex molecules, using DNA strands and their interactions to tether reactants together so they can meet and react in the right order and in the right orientation. DNA assemblers can be designed to evolve autonomously according to DNA-encoded programs.[19] While it is still a long way to industrial-scale production of chemicals using DNA or other molecular assemblers, the path to achieving it has been found, and will be traversed at increasing speed in the coming years.

NANOTECHNOLOGY FOR TARGETED DELIVERY OF DRUGS

Even when a drug is effective, it is very difficult to bring it to the right place. In fact, most of any drug's side effects arise from the diversion of the drug to places in the body where it is not needed. Let's take insulin as an example. Insulin is a hormone that helps to control the amount of glucose in the blood. In healthy individuals, insulin is secreted by the beta cells in the pancreas when sugar levels in the blood rise. The main target of insulin is the liver, which is able to remove glucose from the blood and turn it into glycogen (or fat, or both). The insulin released from the pancreas also acts on adipose and muscle tissue to stimulate glucose uptake. Diabetic patients cannot produce enough insulin (type 1), or their cells become immune to insulin (type 2). In both cases, a usual medical strategy is to monitor glucose levels (which still requires a blood test), and the common treatment is to administer insulin. Despite the efforts of pharmaceutical companies, it is very difficult to create an oral insulin pill that can be effective, since the intestinal lining is a barrier that is very difficult to cross; and so, painfully, insulin remains in injectable form. The chief complication is that most of the injected insulin does not reach the liver, and this means that patients treated with insulin develop many side effects, including weight gain, microvascular disease, renal failure, cardiovascular complications, blindness, and problems with healing—even amputation of limbs. The International Diabetes Federation estimates that 382 million people—8.3 percent of all adults in the world—have diabetes, including 175 million adults who have the disease but are undiagnosed. The numbers will continue to rise, reaching 592 million in a generation, and injecting insulin is still the main, albeit unsatisfactory, treatment.

Another case is cancer chemotherapy. Most of us have been close to someone who has had to undergo the excruciating ordeal of "chemo." The reason for such a harsh treatment is that in order to eliminate cancer cells, the drugs have to reach them in effective concentrations. Unfortunately, this can mainly be achieved by increasing the dose, which has the side effects that have made the treatment so feared: pain, diarrhea, vomiting, blood disorders, nervous system damage, organ damage, infertility, cognitive dysfunction, hair loss.

Frequently, the drug cannot reach the target at all. A particularly tough case is the brain, which is protected by the almost impenetrable blood-brain barrier. This makes treatment of brain diseases very problematic from a pharmacological point of view.

Drug delivery is a major challenge of our time; fortunately, it is coming into the sights of the multidisciplinary approaches that are transforming biomedical research. In particular, nanostructured materials with tailored physical and chemical properties are being investigated for ingenious solutions to the problems of selectivity and access. A minute nanoparticle has, in principle, enough space on it and in it to engineer in several properties that could not be combined in one smaller molecule. For example, nanostructures have the potential to target more than one receptor, extend time of availability in the blood, and/or tackle physical obstacles such as the intestinal lining or the blood-brain barrier. A formidable array of problems has to be overcome: aggregation of the particles on contact with blood or other fluids; their accumulation in the wrong places; their chemical changes inside a living body; the immune system's responses to them; the ability of the kidney, the liver, and immune system cells to get rid of foreign material of any size or shape; and their potential toxicity.

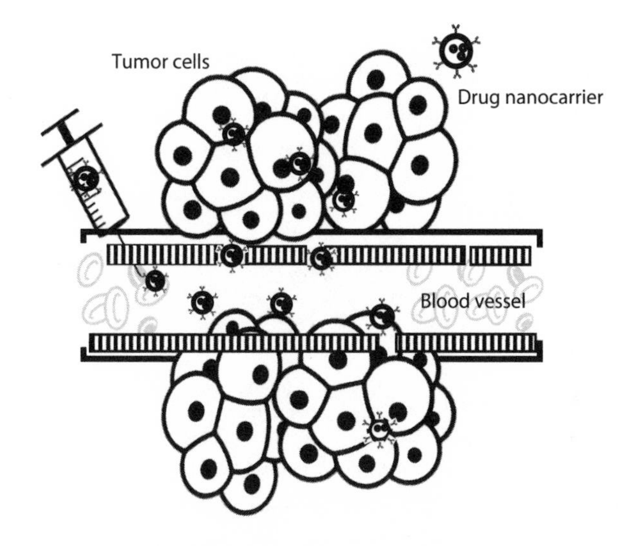

Figure 3.1. Nanocarriers of drugs. *Passive tissue targeting* is achieved by extravasation of nanoparticles through enhanced permeability of the tumor vasculature (EPR effect). *Active cellular targeting* is achieved by functionalizing the surface of nanoparticles with ligands that promote cell-specific recognition and binding. The nanoparticles can (i) release their contents in close proximity to the target cells; (ii) attach to the membrane of the cell and act as an extracellular sustained-release drug depot; or (iii) internalize into the cell.

A particularly interesting finding is that nanoparticles seem to accumulate in tumor tissues much more than they do in normal tissues. This discovery, reported from Japan in 1986,[20] has become perhaps the most cited in the field of nanomedicine. It is the legendary *enhanced permeability and retention effect*, or EPR. In order to grow, tumors generate a network of blood vessels around them (a process called *angiogenesis*); tumors' vessels, however, are usually abnormal. The cells in their lining are not well aligned, and have little imperfections and holes through which nanoparticles can exit the vessel (extravasate) and accumulate in the tumor tissue.

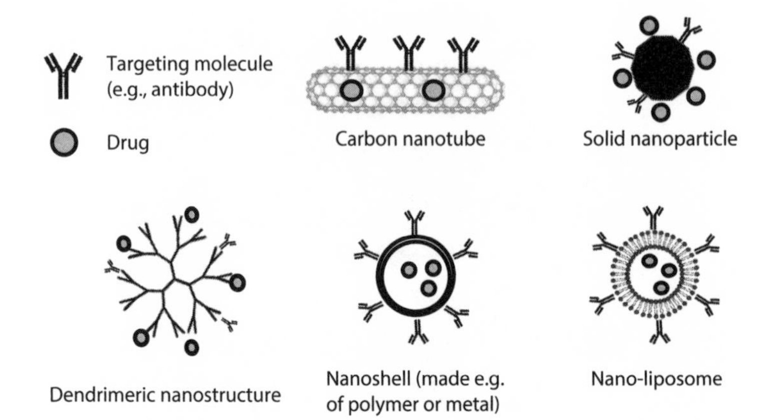

Figure 3.2. A whole range of drug delivery systems have been developed. They typically include a nanocarrier (nanoparticle, nanoshell, dendrimer, liposome, nanotube, etc.); a targeting molecule attached to the nanocarrier; and a cargo (e.g., chemotherapy drugs).

The possibility of engineering nanoparticles that selectively detect and destroy cancer cells using the EPR effect, among other, more sophisticated strategies, has led researchers to engineer myriad nanoparticle designs (generally derived from lipid molecules, polymers, or metals) of specific sizes, shapes, and surface chemical and physical properties, programmed with a fantastic diversity of biological and medical functions. Nanoparticles can be loaded with drugs to achieve a more targeted and concentrated delivery of therapeutic compounds. If biodegradable particles are used, they can also keep up a sustained delivery activity over a longer time. Several systems have been demonstrated, such as polymer-based nanostructures that can self-assemble to target cancer, and nanoparticles made of metals or other inorganic materials that home in on tumors and, when irradiated with light or exposed to magnetic fields, produce enough heat to eradicate them. Other nanoparticles carry one or more contrast agents so that they can be monitored by MRI and, when they reach their target, release drugs via a trigger, such as pH or enzymatic catalysis. Nanoparticles can display a combination of diagnostic and therapeutic agents, currently known as *theranostics*. Molecular assembly techniques can also be used to build nanosystems with multiple functions, such as targeting tumors and facilitating clearance from the body.

Many studies have shown that nanoparticles can detect and kill cancer cells in vitro—in cell cultures in the lab—and in vivo, in mouse models. Very few of these nanoparticles have been translated to clinical use, however—despite much academic effort and the hot pursuit by startup companies all over the world of the "magic bullet" to destroy tumors. In 2016 a controversial article appeared in *Nature Reviews Materials*, examining all the available literature to try to understand why all that research and promise was not being translated into medicine. The problem seems to be that the nanoparticles, in most cases, do not reach the actual tumor. The majority of injected nanoparticles come to a halt in the liver, spleen and kidneys;[21] the body is doing its job—it is these organs' role to clear foreign substances and poisons from the blood. This suggests that researchers may have to control the interactions of those organs using nanoparticles. It is conceivable that an optimal particle surface chemistry, size, or shape is required to access each organ or tissue. One possible strategy is to engineer nanoparticles that can dynamically respond to conditions in the body by altering their surfaces or other properties, much as proteins do in nature. This might prevent them from being filtered out by the liver.

Ultimately, though, more knowledge of the biology and the physics of the delivery process is needed to crack the problem, rather than just changing characteristics of the nanoparticles themselves. Nanomedicine is failing just where more-traditional pharmacology is also failing: by trying to go fast, nanomedicine researchers are overlooking the actual physics and chemistry in play at the surfaces of nanoparticles and proteins in the body. Perhaps, to achieve the goal, the community needs to join forces to coordinate a long-term strategy based on a broader, more contextual view of the problem of drug delivery. As has been shown again and again, tough biological problems need multidisciplinary approaches and coordinated efforts. Collaboration can go hand in hand with constructive ways of organizing competition, as witness the very successful contests established by and for protein folders, mentioned in chapter 2.

That we cannot yet control nanoparticle transport inside the body presents a major limitation for using nanotechnology to diagnose and treat cancer—or diabetes, cardiovascular disease, or any of the other ills for which nanotechnology has raised high hopes and large research budgets. To make further progress, researchers will need to understand much more about the multiscale interactions between nanoparticles and the body's organs and tissues than they do today.

As fast as engineers can invent them, the multidisciplinary approach is bringing together the powers of new databases; advanced imaging and biosensing technologies, to track the movements and interactions of nanoparticles with cells, tissues and organs; and the best computational techniques, from big-data analysis to artificial intelligence algorithms based on sound quantitative mechanistic models. All these, and their explosive synergy, will be required to achieve this breakthrough, but above all, good strategy and good leaders will be the key to success.

The good news is that the very limitations of nanoparticles for cancer drug delivery become advantages on one of the most promising battlefronts against the disease: cancer *immunotherapy*.

NANOTECHNOLOGY TO ENHANCE CANCER IMMUNOTHERAPY

For the last two years, newspapers have been astounding readers with news of miraculous cancer cures. They are based on immunotherapy, a new way of dealing with cancer and other diseases that seeks to unleash and enhance the power of the patient's own immune system to detect and destroy malignant cells. It is without doubt the most promising new cancer treatment approach since the development of the first chemotherapies in the late 1940s. Immunotherapy is bringing about a medical revolution: it is estimated that more than half of current cancer clinical trials include some form of immunotherapy.

Although these results are new, the ideas behind them are not entirely original. Like most stories in modern medicine, immunotherapy's narrative started with bacteria in the late nineteenth century.[22] In 1890, William Coley, an American bone surgeon and cancer researcher, tormented by the death of one his patients from sarcoma (a kind of cancer that affects soft tissues), came across the case of a sarcoma patient whose cancer had disappeared after a bacterial skin infection. This was some twenty years after the discoveries of Pasteur and Koch about microbes and infection. Coley found documented examples going back hundreds of years of people having a "spontaneous regression" of cancer after an infection. Based on his own findings that seemed to be confirmed by these early discoveries, he constructed the hypothesis that infections triggered some kind of immune response in the body that improved cancer survival outcomes, Based on this hypothesis, he developed a startling treatment method: he would infect cancer patients with *Streptococcus pyogenes*—and in many cases he achieved their recovery. Over the years he perfected his method; eventually, he managed to achieve a safer form of the treatment by realizing that you did not need whole living bacteria to produce the desired effect on the immune system. Bacterial proteins extracted by heat-killing the bacteria were as effective as living bacteria to clear cancer. This made it possible to avoid the potentially dangerous infection step that is necessary when living bacteria are used. Furthermore he found out that adding a second bacterial species to the treatment further improved the outcome. He treated hundreds of people. In 1999, a scientific investigation compared his results with those of modern therapies. *His results were better*. Fifty percent of Coley's sarcoma patients lived for ten years, compared with the 38 percent achieved by modern therapies. Coley also got better ten-year survival rates for ovarian cancer and kidney cancer.

The main problem with Coley's work was that it was difficult to standardize at that time. Coley was good at it, but training other doctors to do it was not always successful, mainly because the underlying biological mechanism was not known. When, in the mid-1900s, chemotherapy and radiation therapy were established as the standard of cancer care, Coley's approach was widely forgotten.[23]

The past few decades have seen an outpouring of research on the immune system, yielding a deeper understanding of how cancer progresses and inspiring new ways to stop it. A range of unprecedentedly successful cancer therapies that have recently been reported in the press exploit the complex interactions between tumors and immune cells revealed by this effort. As in most current biomedical research, the progress is the product of reciprocal learning: as soon as the immune system teaches researchers about its intricacies, researchers find tricks for teaching immune cells how to use those abilities more effectively. By using what they learn about the immune system, scientists can assist the immune system to target disease with sharper aim.

In a modern version of Coley's strategy, immunologist Karolina Palucka helped to treat Ralph Steinman's pancreatic cancer with a vaccine based on dendritic cells—the cells he had co-discovered,[24] earning him a Nobel Prize that was announced three days after his death.[25] We now know that immunity is an intricate interplay between the adaptive immune system (which is specific for each *antigen*, or molecule capable of inducing an immune response in the host) and the innate immune system (which is not antigen specific). B cells and T cells of the adaptive immune system have receptors that recognize antigens in a highly specific way. Dendritic cells are the link between the innate and adaptive immune responses. To create an immune response to cancer, the organism needs dendritic cells to present cancer antigens to T cells. But cancers can create an environment that inhibits T cells. The aim of dendritic cell vaccination is to boost cancer-specific T cells that can not only fight existing cancer, but also induce immunological memory to control the reappearance of cancer.

To treat Steinman, Palucka took cells that are precursors of dendritic cells from his blood, activated them in the lab, loaded them with tumor antigens and adjuvants,[26] and then injected the cells back into him, hoping that these cells would activate tumor-specific T cells. Steinman received eight injections of this vaccine over the course of eight months, in combination with chemotherapy. He survived for 4.5 years after his diagnosis—longer than 95 percent of patients with this disease.

Some of the results reported are staggering—patients in terminal stages of the disease having a complete recovery. But many other patients' cancers don't respond, and researchers struggle to understand why.[27] Most immunotherapy treatments, especially those directed against solid tumors, have so far benefited only a minority of patients.

Recent research is showing that nanotechnology might help to overcome some of the limitations of current immunotherapy strategies.[28] For example, in order to get a robust antitumor response, it is important to expose each dendritic cell to both antibodies and *adjuvants*, "booster" substances that enhance the immune response. Currently both are administered as separate drugs, and this has the disadvantage that some dendritic cells only get exposed to one or the other of them. Co-encapsulation of both antigen and adjuvant in a single nanoparticle enables co-delivery to the same dendritic cell, which leads to a more effective activation of the T cells. It is also possible to encapsulate both antigen and adjuvant in a particle that acts as a reservoir and is able to release the molecules over a longer period of time; this could be useful to avoid booster injections and also to improve the efficiency of the treatment.

While vigorous T cell responses have often been induced in mice in the lab, this has been difficult to achieve in large mammals. But preliminary trials of nanoparticle-based vaccines in nonhuman primates and humans show results very similar to those in mice, which suggests that nanoparticle-based vaccines could become an important breakthrough.[29]

Targeting nanoparticles to dendritic cells is much easier than targeting them to cancer cells. Because the particles resemble pathogens in size, immune system cells such as dendritic cells readily take them up, phagocytosing a nanoparticle in the same way as they would a virus.

Nanoparticles also offer another advantage; they naturally accumulate in the spleen (and other secondary organs of the lymphatic system). This is a problem for targeting a specific tumor, but it is an advantage in immunotherapy, because there are many dendritic cells in the spleen. Furthermore, organs like the spleen, unlike solid tumors, do not present a complex physical barrier that prevents nanoparticles from reaching the target cells. Using nanoparticles, it is also possible to target multiple types of dendritic cells by selecting different types of antibodies; in this way it is possible to induce both cellular and *humoral*[30] immune responses. By prolonging their retention at the tumor site, nanoparticles can also be used to produce an immune response in the local environment of the tumor. This reduces toxicity to the patient and allows for strong doses of extremely potent immunostimulatory molecules while still promoting a response in the whole body.

Scientists at MIT managed to stick adjuvant-loaded nanoparticles to the surface of T cells to create "pharmacytes," producing much better results in tumor elimination by T cells.[31] This approach has been taken further: immobilizing nanoparticles loaded with immunostimulatory drug combinations to the surface of adoptively transferred T cells (tumor-specific T cells extracted, expanded in the lab, and injected back into the host) has induced an immune response in tumors of mice superior to that obtained by loading the drugs separately. These drugs can help T cells override the immunosuppressive signals that tumors emit to shield themselves while they grow and spread, but, given systemically, these drugs are very toxic and not very effective. By concentrating them at the surface of T cells, the authors demonstrate that a nanoparticle-based strategy can by much more effective.[32]

Wenbin Lin's lab has designed nanoparticles between 20 and 40 nanometers in size, loaded with light-sensitive chemicals and coated with a special shell that allows them to survive a long time in the bloodstream and find tumors. Once the particles infiltrate their targets, shining near infrared light on the tumors triggers a chemical reaction that ruptures the cells. The broken tumor cells expose their antigens to dendritic cells, enabling a powerful anti-tumor response by the T cells nearby. These nanoparticles have been applied successfully in mouse models of colon cancer and, more recently, breast cancer—a type of cancer that often does not respond to current immunotherapies. The immune response triggered by these nanoparticles is not limited to the primary tumor site: metastases were also wiped out.[33]

NANOPARTICLES FOR GENE EDITING AND GENE DELIVERY

Gene editing using CRISPR has also been making headlines in newspapers all over the world. In this case, too, the tools for medical research come from bacteria. CRISPR (pronounced "crisper") are bits of DNA that bacteria use to flag the beginning and the end of any DNA segment that does not belong to the bacterium but that has been inserted in its genome by, for example, a viral attacker. The Cas enzymes of the bacterium recognize the CRISPR DNA flags, attach to them, and cut the exogenous DNA elements at both ends, effectively editing the foreign genetic information out and acting as a tool of the bacterium's immune system.

The CRISPR/Cas system is a game changer in biological sciences and biomedicine, because it can be used in other organisms to cut and replace specific parts of DNA using Cas proteins. The potential of this technique is enormous; among many other things, it can lead to the correction of genetic diseases, as well as to pest-resistant crops and malaria-free mosquitoes.

The success of CRISPR/Cas–based gene editing relies, like most of the therapies I have discussed, on the efficient delivery of the therapeutic material to the part of the body where it needs to act. Otherwise, the treatment is either inefficient or toxic, or both. In this case the cargo must be delivered to the inside of the cell, which is not an easy task. The cell usually isolates foreign material inside itself by encapsulating it in a lipid membrane vesicle, the so-called *endosome*. Endosomes are impermeable, and drugs trapped inside them are useless. Breaking the endosome is a fundamental goal of all the therapies that target molecules inside the cell. Groups are already working toward improving the effectiveness of gene editing by designing lipid nanoparticles that can cross the mammalian cell membrane to effectively deliver Cas proteins to the intracellular environment.

Apart from editing genes, nanoparticles are also being investigated for other forms of gene delivery. Almost any sequence in the genome, and every coding and noncoding messenger RNA in the cell, has the potential to be become a therapeutic target that can be addressed by an artificial nucleic acid. Over the past two decades, much research has been devoted to the clinical application of gene-based therapies for treating or preventing a wide range of diseases. In clinical trials, however, success has been very limited. Most of this research has used viruses to deliver the gene therapy material. Viruses are able to insert their DNA into the host's cell nucleus and edit its genome, and so hacking the machinery of viruses has been pursued as a possible strategy to achieve beneficial gene editing, but so far it has led to incomplete results, unwanted side effects, and complications. Nanoparticles have the potential to overcome some of the limitations of viruses for gene delivery.[34] Several approaches are currently being developed, including libraries of biodegradable polymers[35] that can bind DNA and condense it into small, stable, nontoxic nanoparticles, able to cross the cell membrane and become active in the intracellular medium. Although effective treatments have yet to be achieved, such nanoparticles have proven useful to efficiently deliver gene therapy in a mouse model of cystic fibrosis. Using a nanoparticle-based formulation that can be delivered through the nose, researchers have demonstrated efficient gene delivery to the lungs of mice, which is a very promising advance.[36] Success will probably come by combining what we've learned from biological research with viruses and from nanoparticle delivery of cargoes inside the cell.

CONTROLLED RELEASE OF DRUGS AND MOLECULES FROM POLYMERIC MATERIALS

Judah Folkman, the youngest full professor at Harvard Medical School at the age of thirty-four, reported in 1971 that solid tumors depend for their survival on their ability to generate blood vessels around them.[37] He hypothesized the existence of growth factors secreted by tumors to promote the growth of blood vessels; after he experimentally found them, he proposed blocking the growth factors as a strategy to kill tumors, or at least to halt their development. Folkman pioneered the collaboration between the university and industry in medical research, and in 1974, his medical inventions were the first for which Harvard allowed its faculty to submit a patent application.

A decisive moment for the field of drug delivery came when Robert Langer, a chemical engineer, joined Folkman's lab as a postdoc. After extensive trial-and-error experimentation, Langer developed a polymer system that provided slow and constant release of an anti-angiogenesis factor. Furthermore, he found polymers that could be used for controlled release of molecules of virtually any molecular weight.[38] Langer's results were met with great skepticism by the scientific community—and also by the patent office. Langer later tinkered with the fabrication procedure so that the polymers could be made into microparticles of different shapes and sizes, which could be implanted to continuously release biomolecules for over 100 days at a constant rate. This discovery was initially ridiculed by scientists, and Langer's first nine grant applications were turned down. He almost lost his professorship.

He persisted. Over time, his science and technology enabled the medical use of many peptides, nucleic acids, sugars, charged low molecular weight pharmaceuticals, and proteins. These larger molecules have extremely short lives (minutes, in some cases) in the body, which deploys its immune machinery to destroy them. Controlled, localized delivery is required if the molecule is to reach its target. The polymeric systems developed by Langer have had a significant impact on developmental biology and on the deployment of many clinically used therapeutics.[39] Langer is currently revered as the most important figure in the field of drug delivery.

In 1969, a controlled release system based on Langer's polymeric microspheres was approved by the FDA for use with a large-molecule peptide drug to treat cancer. This drug delivery system has become the most widely used for advanced prostate cancer patients. Later, the polymeric microspheres were optimized for the treatment of Type 2 diabetes, alcoholism, schizophrenia, and narcotic addiction, and they became implantable materials to release high levels of chemotherapy agents to brain tumors in a sustained, controlled way.

In 1996, for the first time in more than twenty years, the FDA approved a new treatment for brain cancer, using Langer's microparticles; it was also the first time that the FDA had approved a system that directly delivered chemotherapy to a tumor. Its use has extended the life of patients, from a few weeks to, in some cases, many years. Over 2 billion lives have been improved worldwide by the technologies developed by Langer and his lab over the years.

Langer also made the first "smart polymers" that could be controlled with external signals—for example, ultrasound or magnetic fields—to release drugs on demand. Langer has also developed drug delivery systems that can be activated by enzymatic activity, or even by intelligent chemical microchips.[40]

Microchip-enabled implantable pumps have been used to deliver human parathyroid hormone fragments in clinical studies for the treatment of osteoporosis. A bidirectional wireless communication link with the implant was used to program the dosing schedule and to monitor the status of the implant during the operation.[41]

CONTROLLED RELEASE OF DRUGS FROM SKIN PATCHES USING BIORESPONSIVE MATERIALS

Improved knowledge of the responsive mechanisms of biological systems is leading to innovations in materials chemistry, biomolecular engineering, pharmaceutical science, and micro- and nanofabrication. Together, these advances are facilitating the creation of bioresponsive materials for a range of applications, including controlled drug delivery, diagnostics, tissue engineering, and biomedical devices.[42] This field seeks to evolve from the pioneering ideas of Langer toward smarter release of drugs when and where they are needed.

Several prototype materials and devices are being fabricated to respond to specific biological, chemical, and physical environments. A skin patch displaying 121 microneedles, loaded with nanoparticles designed to release insulin when glucose levels are low, was reported in 2015.[43] The authors were inspired by the function of the pancreatic beta cells, which tightly control insulin levels in the body by monitoring blood glucose concentrations (plate 9).

There is increasing interest in developing skin patches for monitoring and treating disease. Diabetes is usually the model system in which to test and develop the technology. In 2016 a collaboration of several Korean institutions reported a flexible graphene-based device for sweat-based diabetes monitoring. The patch incorporates a heater; temperature, humidity, glucose, and pH sensors; and polymeric microneedles that can be thermally activated to deliver drugs through the skin.[44] Electronics, materials chemistry, and biology merging to monitor and treat disease is the trend of many of the cutting-edge programs in this field.

IMPLANTS FOR IMPROVED IMMUNOTHERAPIES

A particularly interesting use of porous materials is their adoption in the framework of immunotherapies. Implants the size of a pill are currently being designed to recruit and stimulate immune cells to attack tumor cells.[45] The implants contain molecules that recruit dendritic cells and DNA fragments, which then stimulate the dendritic cells by mimicking bacterial infections. The ingenious plan is to include ground-up tumor tissue from biopsies to teach dendritic cells that these are foreign substances that must be eliminated.

The implants have been a complete success in experiments treating melanoma in mice. The WDVAX vaccine is currently being tested in patients with metastatic melanoma in a phase 1 clinical trial. This approach is currently being extended to other

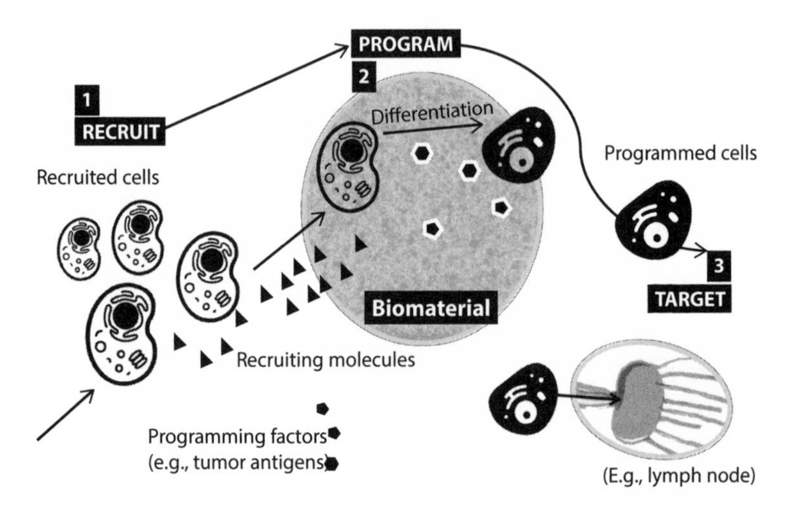

Figure 3.3. General strategy to use biomaterial systems for immune therapies, to recruit and reprogram cells to target tumors.

types of cancers, such as brain gliomas and breast cancers. Similar approaches are being tested on lymphomas.[46]

Because vaccine implants require minor surgery, research is under way to make the strategy less invasive. For example, a method has been reported that would swap out the polymer scaffold for a cryogel of porous microparticles;[47] following injection under the skin, these particles clump together to form a sort of immune-priming depot.

TOWARD THE SUPER-ENHANCED IMMUNE SYSTEM

No matter how much power a drug, enzyme, or cell has to cure a disease, the crucial feature needed for therapeutic success is to deliver it to the right place. The body has many barriers to prevent this: the mucosa of the lungs, the blood-brain barrier protecting the brain, the filters of the kidneys, and the intestinal environment are all designed to prevent pathogens and poisonous molecules from killing the organism, and so they also prevent drugs and treatments from reaching the targeted areas. Cancers and tumors have evolved to prevent any external agents, such as immune cells, molecules, or proteins, from attacking them. These barriers are physical as well as chemical, and much of the research in biomedicine is devoted to finding strategies for crossing these barriers and delivering treatment where it is needed, while avoiding toxicity. Nanomaterials will be key to achieving this goal because on them it is possible to engineer the right chemistry and the right physics to achieve it. Nevertheless, the task is far from completed.

A future area of drug delivery that can be improved by nanotechnology is to extend the targeting of drugs to the interior of the cell. Currently, most drug targets are situated in the cell membrane. Being able to cross the cell membrane to deliver drugs, nucleic acids such as DNA and RNA, and short proteins to intracellular targets would open up the possibility for many new treatments. While this is being increasingly achieved, it remains a fundamental hurdle, especially for gene-based therapies.

The general self-assessment of the field of nanomedicine that is pouring into the scientific literature is that we are getting closer, but we are not quite there yet. The results so far suggest that the most effective way to progress will be to develop common strategies in the research community. Driven and inspired women and men will be needed as leaders to push for those common goals and strategies. My impression is that the nanomedicine community is finally bringing the necessary focus to bear on the toughest problems: targeted delivery to different parts of the body, and intracellular delivery. The speed of research on new materials, new biology, new techniques, and quantitative methods of monitoring disease will contribute to finding solutions. The history of medicine indicates that our urge to solve problems is unlikely to disappear.

Another key to achieving success will be to develop better mathematical modeling of drug delivery to help design and refine treatments for maximum efficiency and results. Nanotechnology-enabled biosensing platforms will increasingly be able to monitor molecules and a variety of physical and chemical signals in patients in real time. Smart mathematical modeling techniques currently in development include AI algorithms to make sense of the large data sets specific to individual patients. From these compiled data the algorithms will sculpt mechanistic models of disease and treatment that can predict a new patient's optimal dose and response, based on actual physical processes within the body across the temporal and spatial scales.

The convergence of nanotechnology, electronics, physics, mathematical modeling, artificial intelligence, materials science, and biology will result in ever smarter drugs, delivery systems, and diagnostic strategies, with the ultimate goal of continually monitoring the body to be able to respond quickly and fix it when it first malfunctions. This is, in fact, what our own immune system does: continuous monitoring and repair. From this perspective, it is not surprising that the technologies and sciences of health are now converging in immunotherapies. Our immune systems have weak points, which cancer and pathogens can exploit. Our eagerness to heal forces us to learn how biology does it, and to identify the weak points so we can eliminate them. In the future, we will probably be able to create super-enhanced immune systems, boosting evolution's already astonishing creation with hybrid transmaterials and smart mathematical models that take it to the next level. That will bring us closer to the ultimate goal of all this striving: gradual mastery of the basic principles of biology, with its ability to create and to heal.

In the next chapter, I will show how these ideas are all being deployed in one of the most potentially transformative fields of research: regenerative medicine. Whether trying to recreate living tissues in the lab for transplantation or to enhance the regenerative powers of the body, researchers in this area are quickly integrating all the knowledge that I have introduced in this and the previous chapters—from bioelectricity and the mechanics of biology to nanotechnology, biosensing, and mathematical modeling.

4

RECREATING TISSUES AND ORGANS

Going through life keeping our physical integrity unscathed is no cakewalk. Cancer, accidents, violent encounters, medical treatments, diabetes, infections, and burns often lead to severe tissue damage and organ failure. Fear of maiming and amputation is as significant to our human identity as is our wholeness. From ancient times, humans have dreamed of being able to fix and replace diseased, diminished, or lifeless body parts, and even to regenerate the whole body to achieve immortality.[1] This is a yearning that has never been suppressed, partly because humans have always witnessed salamanders, axolotls, starfish, crayfish, newts, hydras, and worms regrowing whole arms, legs, tails, jaws, internal organs, and eyes in front of our jealous, admiring gaze. Humans can regenerate skin if the injuries are not too big; the liver, kidneys, and bone can also regrow to some extent. Women regenerate our endometrium during each menstrual cycle. But for the most part, evolution has hidden the recipe for the regeneration of human organs and limbs in our ability to do science.

Currently, the main remedy is transplantation. This is a difficult, expensive, and culturally loaded activity that can only be performed by very skilled surgeons, who have become the stuff of hero worship and legend.[2] Tragically, the demand for organs and the increase in the number of doctors and institutions that can successfully carry out transplants has led in some cases to shocking organ market activities. Organs are routinely and inhumanely removed from the weakest people of this world to feed the survival demands of those who can afford them, profiting from desperation.

In this chapter, I will very briefly review the history of scientific and medical understanding of healing and tissue regeneration, from the discovery of the unit of life, the cell, in the seventeenth century to our current accomplishments in isolating and generating stem cells. The efforts to regenerate and repair organs have given rise to the field of tissue engineering, a multidisciplinary ground where medicine, evolutionary and developmental biology, physics, engineering, and nanotechnology all meet to solve fundamental questions about how cells communicate with each other and with their environment. These questions are expediting the development of new nanomaterials, sensors and monitors of biological and physical activity in vivo, new mathematical models that capture the complexity of living tissues, and new strategies such as the control of biological electricity to regenerate organs. As the first artificial organs have already been successfully transplanted, the field transcends medicine and leads to profound questions about the identities of cells. How does a cell know that it has to behave like a heart cell or like a neuron, or when it needs to die? How does a body or an organ know its shape? Can we control the behavior of cells with artificial materials or electrical and mechanical signals to create post-evolutionary functionalities that do not exist in nature? Can novel biohybrid structures be fabricated that combine the engineering principles of biology and nanotechnology to create new transmaterial devices—or even creatures?

FROM THE DISCOVERY OF CELLS TO STEM CELLS

As you will recall from chapter 1, Robert Hooke in England managed to construct a microscope that allowed him to observe that plants were made of *cells*, a term he coined to describe the plants' building blocks (because their packing reminded him of the small rooms in monasteries where monks lived). He published his findings in the book *Micrographia*, in 1665. More than a century later, Lorenz Oken, a German biologist, used the observations of microscopists to develop hypotheses about the cell's roles in life. In 1805 he stated that "all organic beings originate from and consist of vesicles or cells." At that point science got on the right track to understand tissues.

Microscopy that could reveal cells in tissues led to progressive observation of the stages of development of multicellular organisms as they grow in complexity from a small number of cells. Science learned that in sexual organisms, the fertilized egg divides to form a ball or sheet of similar cells, the blastula or blastoderm; the embryonic *pluripotent stem cells* contained in the blastula then go on to divide and *differentiate* to create the diversity of tissues and organs and specialized cells that can be found in nature, from leaves to eyes and brains. To the early observers of cellular life, it soon became apparent that if we learned how to get hold of those pluripotent stem cells, and how to activate and assemble them in the right way, we could perhaps build tissues and replace, even recycle, our malfunctioning organs.

Jacques Loeb[3] in the nineteenth century was probably the first person who thought that learning biology's tricks for creating tissues was possible, and had the audacity to try. Loeb was an iconoclastic German Jewish physiologist who developed "an engineering standpoint" on biology, influenced by the practical ideas of the botanist Julius Sachs.[4] Like Sachs, Loeb was interested in *controlling* biology; his experiments and ideas led him to establish a long correspondence relationship with the physicist-turned-philosopher Ernst Mach. Mach was an advocate of "scientific positivism": he thought that scientific research had to be considered as part of its social and cultural context, as a way to human social and economic advancement, in a deep relationship to technology. For Loeb, reinforced by Mach, a biologist should be an engineer.

In a series of pioneering experiments with tubularians (small animals related to jellyfish) in the Naples Zoological Station, he began to study regeneration as a capacity of organisms to react to changing environmental conditions. In a letter to Mach in 1890, he wrote, "The idea is now hovering before me that man himself can act as a creator even in living nature, forming it eventually according to his will. Man can at least succeed in a technology of living substance."[5] He fervently pursued his ideas at the University of Chicago and became a celebrated scientist; he inspired the protagonist of *Arrowsmith* (1925), the first novel to romanticize the world and culture of science, written by Sinclair Lewis. In 1897, Loeb first reported the idea of growing cells outside the human body, in "cell cultures," and in 1914 he described the activation of sea urchin eggs to begin embryonic development in the absence of sperm, using ultraviolet rays.[6] Ross Harrison was the first to grow cells in vitro in 1907, developing the first neuronal tissue culture line from frog embryo ectoderm.[7] Throughout the twentieth century progress in growing and maintaining cell cultures continued, increasing the ability to grow cells from specific tissues in vitro. A fundamental piece of this story is John Enders, who found a way to culture human embryonic cells (using fetal tissue from abortions and miscarriages) for the development of viral vaccines in 1952.

The interest in cell types, cancer, development, and cell cultures led scientists in the 1950s to investigate the biology of "monster tumors," or *teratocarcinomas* (malignant) and *teratomas* (benign). Teratocarcinomas and teratomas have terrified and intrigued people since antiquity, because they are composed of a slapdash mixture of tissues and organs and often contain teeth, bone, hair, muscle, skin, and even eyes, all in a single tumor. Teratomas are very rare, especially in mice. This meant that although potentially a boon for understanding development, teratomas were very difficult to study without an accessible animal model. This changed in 1954, when the first mouse strain was reported that showed an incidence of spontaneous testicular teratoma of about 1 percent.[8] The research on teratocarcinomas grew in the 1970s with the tide of interest in mammalian developmental biology and cell differentiation.

In 1970, retransplantable teratocarcinomas from mouse embryos grafted to extrauterine tissues confirmed that stem cells in teratocarcinomas were essentially embryonic; this explained their pluripotency, i.e., that they could turn into almost any tissue inside the tumor. Between 1970 and 1974, several isolations of cell lines derived from teratocarcinomas were reported.[9] Improvement of cell culture techniques made it possible to reliably clone cultures of mouse pluripotent teratocarcinoma cell lines.[10] Following the work with mice, the isolation of human carcinoma cell lines was reported in several publications between 1977 and 1980. These cells were useful for understanding tumors and cell differentiation, but had limited medical utility because of their cancerous origin. For medical applications, healthy human stem cells were needed.

The isolation of mouse embryonic stem cells was reported in the early 1980s, but the derivation of the first human embryonic stem cell and germ cell lines took longer. Besides the technical and scientific hurdles, most investigators were understandably reluctant to work in a field that is fraught with ethical dilemmas, political difficulties, and even legal dangers. The ban in the United States on supporting human stem cell research with federal funds put an important brake on progress.

The breakthrough came in 1998, when James Thompson visited Israel at the invitation of Dr. Joseph Itskovitz of the Technion, the Israel Institute of Technology. Their collaboration led to the first successful isolation of human embryonic stem cells.[11] The cells were isolated from embryos donated by couples who had undergone fertility treatment in Haifa. Once the first cell lines were isolated, scientists and the public realized that stem cells could have a colossal impact on medical practice. At the turn of the century, the pages of the press began to be inundated with articles on the promise of the science of stem cells.

The work progressed at a remarkable speed, and in the United States the federal funding ban gradually relaxed, especially under the Obama administration. In 2012, Shinja Yamanaka of Japan and Sir John Gurdon of the UK shared the Nobel Prize in Physiology or Medicine for the discovery that mature, differentiated cells can be reprogrammed to become pluripotent. This finding means that pluripotent stem cells can now be created from the cells of any tissue (skin, for example) without the need of an embryo, which clears many ethical and legal dilemmas and also opens the way for self-transplantation. In 1962, Gurdon had done pioneering experiments removing the nucleus of a fertilized egg cell from a frog and replacing it with the nucleus of a cell from the intestine of a tadpole. This modified egg cell grew into a new frog, proving that the mature cell's nucleus still contained the genetic information needed to form all types of cells. Before this experiment, most scientists had thought that once a cell was already differentiated, it could not return to the pluripotent stem cell state. In 2006, Yamanaka succeeded in identifying a small number of genes within the genome of mice that he could activate to reprogram skin cells from mice back to the stem cell state, in which they had the capacity to differentiate and grow into any of the types of cells of the body: neurons, heart muscle cells, retinal cells—any cell.

While stem cells have within themselves the potential to become any cell in the body, to create a living differentiated tissue from them—and even more so, a complex solid organ—remains an enormously complicated task.

EARLY TISSUE ENGINEERING

In tissues, cells are assembled in complex, more or less symmetrical three-dimensional structures. In most cases, they are embedded in a complex network of active nanostructured "cables" made predominantly of protein that form a scaffold: the *extracellular matrix* (ECM). The ECM provides a structural, physical, mechanical, and biochemical environment to support the cells' growth and interaction in tissues. The extracellular matrix is an active gel secreted by cells, and is made of a large variety of fibrous proteins and sugars that include collagen, elastin, hyaluronic acid, and proteoglycans (names the cosmetic industry has popularized in its ads, with good reason).

When we incur a wound, the extracellular matrix is destroyed; when cells move into the wound space, the information of the extracellular matrix has disappeared, so the cells generate scar tissue to fill the void. Scar tissue is stiffer, and does not conform well to the rest of the structure. In scars the new collagen secreted by cells has a different structure and is usually more aligned in one direction.

Early on, it was recognized that in order to generate tissues, it would be necessary to recreate the environment that the extracellular matrix provides in tissues, by generating artificial scaffolds in which to seed the cells. W. T. Green at the Boston Children's Hospital was perhaps the first to realize this requirement in the 1970s, when he was trying to work out how to regenerate cartilage by seeding chondrocytes (cartilage cells) onto spicules of bone that he then reimplanted in mice. Although his pioneering efforts were not fully successful, he rightly predicted that with the advent of innovative biocompatible materials, it would become possible to generate functional tissues.[12]

Several years later, also in Boston, the first attempts were made to produce artificial skin in the lab, using a collagen matrix to support the growth of a type of skin cells called *fibroblasts*.[13] This work was followed by breakthroughs in treating burn patients with epidermal sheets of keratinocytes (the main cellular components of skin), alone or atop a dermal layer grown from fibroblasts seeded on collagen gels.[14] These examples represent the foundational stage of the new discipline now known as "tissue engineering."

A momentous day for the field came in the mid-1980s, when Joseph Vacanti of the Boston Children's Hospital approached Robert Langer, the MIT chemical engineer who had started a revolution in drug delivery in 1976 with his application of polymers for controlled release of molecules, as I briefly recounted in chapter 3. Vacanti was interested in designing scaffolds of artificial materials for cell delivery; he thought that such materials would allow the scaffolds to be given predetermined physical and chemical properties, in contrast to the less controllable biological materials (such as collagen) that had been used up till then. The collaboration led to an extensive program to generate functional tissues.

With the development of synthetic biocompatible and biodegradable materials for biomedical applications, research on tissue engineering and biomaterials finally took off in the 1980s. Biomedical engineering departments were established at major universities around the world. The initial successful results were summarized by Langer and Vacanti in the journal *Science* in 1993.[15] These achievements included the fabrication (by Vacanti, his brother Charles, and colleagues) of a scaffold that was implanted under the skin of a mouse to guide bovine cartilage-forming cells to grow tissue in the shape of a human ear.[16] The scaffold was biodegradable: it dissolved slowly while the tissue was growing. The image of the "Vacanti mouse"[17] displaying a human ear on its back is perhaps the most visually disturbing example of tissue engineering, but the startling, groundbreaking experiment paved the way for future applications.

Out of this intensive activity the first commercial products started to appear on the market. Interpore's Pro Osteon coral-derived bone graft material was announced in 1993. In 1996, Integra's Artificial Skin was approved as a nonbiological tissue regeneration product. In 1998, the General and Plastic Surgery Devices Advisory Panel to the U.S. Food and Drug Administration (FDA) recommended unconditional endorsement of Apligraf (Graftskin) Human Skin Equivalent for the treatment of venous leg ulcers. Apligraf, by Organogenesis, was the first manufactured living human organ to be recommended for approval by an advisory panel to the FDA.[18] Ever since then, fundamental science, technology, and commercial application have all run in parallel, often working in partnerships to create better tissues and materials.

ARTIFICIAL MATERIALS TO CONTROL THE FATE OF STEM CELLS

The ability to steer pluripotent stem cells to differentiate into cells of any desired tissue is a fundamental pursuit of the field. Most of the endeavors of the biological and biomedical research communities were initially devoted to finding the molecular signals (e.g., growth factors) that trigger the differentiation. But the irruption of physical scientists and engineers into the field led to a fundamental change in strategy. Forces and mechanics are now recognized as a fundamental piece in the construction of tissues and in the transmission of information from the molecular to the cellular and tissue levels.

As I introduced in chapter 1, cells exert and respond to forces. Furthermore, stem cells can be driven to differentiation by the mechanical properties of the environment where they attach and grow. This opens up the possibility of constructing tissues with biocompatible materials engineered at the nanoscale to transmit physical and chemical signals that guide cell differentiation. A new generation of multidisciplinary tissue engineers is currently working in labs all over the world to bring physics and engineering principles to biology, with the goal of reconstructing living tissues by using artificial material scaffolds that mimic the physical and chemical conditions of real tissues.

Nanotechnology has become a fundamental tool of tissue engineers. As I explained in chapter 1, integrins and other cell adhesion proteins that act as molecular "hands" to attach to the outside world need nanometer-scale handles to hold on to; hence nanotechnology is required to fabricate the right architectures to enable cells to attach, survive, evolve, and divide.[19] Apart from materials, nanotechnology is also needed for measuring. Mechanical properties of living cells and tissues are not easy to measure, especially at the scale of proteins and subcellular structures. AFM and other nanotechnological tools are currently being utilized to quantify the mechanical properties of cells and materials, such as their elasticity, their viscosity, and their poroelasticity (the sum of mechanical properties arising from having a porous structure with water inside).[20]

Having quantitative measurements of the mechanical properties of cells is important for designing the materials of scaffolds, and also for making mathematical and computer models of the mechanical behavior of cells and materials. Computer models can assist in understanding the biology, and, crucially, they have the capacity to become predictive. The ultimate goal is to create the "tissue engineering app" of the future, a computer program that would help to design a tissue construct and a protocol for each application. We are still very far from an app-based design; the most fundamental pieces of information that are needed to put it together are, on the one hand, quantitative measurements, and on the other, good mathematical modeling—and they are both still very much works in progress.

Research has shown so far that to guide the growth of a tissue with an artificial scaffold, the scaffold must match the mechanical, structural, and chemical properties of the tissue one is trying to recreate. Having detailed structural and mechanical measurements of both the healthy tissues and the scaffolds becomes a fundamental part of the research, which is also a work in progress.

NANOSTRUCTURED MATERIALS FOR TISSUE ENGINEERING

Several methods of fabricating nanomaterials are currently being used to produce tissue-engineering constructs. Polymeric fibers that mimic the topography and mechanical properties of the extracellular matrix are being produced, for example, by "spinning" polymers with electric fields (*electrospinning*). The protein nanotechnology researchers introduced in chapter 2 are also producing proteins and protein fragments that can be used to create scaffolds and meshes for culturing cells in three dimensions. In a modern, nanotechnological version of paper, nanostructured cellulose is also being used to create the scaffolding for cells. Methods of producing nanostructured porous materials by freeze-drying solutions of polymers, and of mimicking nature's hydroxyapatite structures (which harden bones and teeth), are also in the works.[18]

Another attractive possibility is to build tissues using building blocks. Microporous polymeric materials with controlled sizes and shapes can be used to construct three-dimensional architectures, or can even be implanted or injected into the body. The blocks can be directly loaded with cells, drugs, and signaling molecules, offering long-term local delivery of therapeutic molecules at the injury site or within the cell culture. This modular approach to tissue engineering has evolved in several very ingenious strategies that aim at the self-assembly or bottom-up selection and arrangement of the building blocks, creating complex tissue architectures that can even include blood vessels and other nutrient transport architectures within the construct.[21] Hagan Bayley's lab at Oxford has invented a system for modular tissue engineering that involves 3-D printing of modules the size of cells with complex 3-D microarchitectures; it is currently being developed by the company OxSyBio.[22] A common strategy that is alluring in its simplicity is to build three-dimensional tissues out of stackable sheets.

Making a fully functional organ using an artificial scaffold is a very hard task. One of the main problems is that diffusion of nutrients and oxygen in the artificial constructs is difficult to achieve. This means that although it is possible to construct small functional cell cultures of a few millimeters, it is difficult to make them bigger and to keep them alive, because oxygen and necessary nutrients do not reach the cells. Transport in real tissues is achieved by a combination of very complex biological and physical mechanisms, including the generation of blood vessels that can carry a constant cargo of oxygen and nutrients to cells in very dense environments. It is known that it is only possible to supply oxygen and nutrients to, and remove wastes from, cells that are less than 200 microns away from a blood capillary. The generation of blood vessels in a fabricated tissue is still a complex biological and bioengineering problem. Sophisticated bioreactors that mimic the conditions of the body are being designed, as well as strategies for overcoming the many diverse challenges. A further complication for making large tissues is finding appropriate cell sources that can deliver sufficient quantities to generate a tissue large enough to have medical relevance. In all circumstances, scientists strive to keep learning by making, advancing the field scientifically and technologically even when the final goal is difficult to reach.

ENGINEERING ORGANS

While fully functional fabrication of the most complex human organs is still a distant aim, great progress is being made in many specific tissues. The simplest organs to fabricate are flat and relatively stiff, such as skin. The next level of complexity is to engineer tubular structures such as blood vessels and tracheas. Hollow non-tubular structures such as the bladder are next in the scale of difficulty. The holy grail of the field is the production of solid organs such as the heart, kidney, and liver.[23]

The area of biomaterials for orthopedics and ophthalmology is particularly active, especially in the tissue regeneration of cornea and cartilage.[24] Articular cartilage and the cornea have many common characteristics: for example, they are not vascularized, which is to say they do not need blood. Both cornea and cartilage have a dense, highly organized extracellular matrix with relatively low cell density. Both tissues suffer from an inability to self-repair, and they form bothersome scars when they are damaged, which can lead to reduced vision, in the case of injured corneas, and weaker mechanical response in cartilage. Furthermore, they are not extremely soft, which is an advantage for creating matching artificial materials. Several new commercial products have been developed.[20] But it is difficult for the new materials to displace the cruder cartilage-substitute implants available on the market since the 1960s. In the case of cartilage, plastic and metal implants are routinely used in the clinic, and although they produce many problems and side effects, they are still preferred because they are well known. The new tissue-engineered strategies are more complex and expensive and still have not delivered clear solutions. A main complication is how to interface cartilage engineered in the lab with the host tissue of the patient. Several strategies have been put forward, including biological adhesives, but it remains challenging to bring the new constructs to the clinic.

Tissue engineering is also a promising technology for repairing spinal cord injuries.[25] Spinal cord injuries often lead to permanent motor and sensory dysfunctions, which have devastating consequences for patients' quality of life. Many scaffold materials are currently being investigated for use in these applications. Neural tissue is particularly soft, and a big challenge is to create a scaffold soft enough to sustain healthy neurons. Injectable, very soft-gelling hydrogels seem to have the best clinical potential, because they can be delivered into the injured region with minimal tissue damage. A particularly exciting development is the use of materials that can electrically stimulate neurons. These materials can deliver an electrical current to cells and thus have great potential for nerve regeneration and functional recovery. They can also be utilized as a guiding channel for broken nerves, to promote nerve regeneration, or to mimic neural tissue for *in vitro* testing. Several clinical trials are under way, studying the effects of electrostimulation as well as stem cell transplantation for spinal cord injury using bioengineered materials.[26] In 2015, a team headquartered at Harvard reported the invention of soft 3-D electronic scaffolds that can be delivered to specific regions, such as cavities and living organs, through a syringe needle.[27] These injectable electronic materials could be used to monitor physiological and electrophysiological signals in vivo, and also to deliver therapeutic electrical signals. The integration of the injectable electronics with other functional units, such as wireless components, is expected to lead toward implantable bioelectronics and continuous biomonitoring that could be connected, e.g., to a smartphone.

The science media often highlight the work of Anthony Atala and his colleagues at the Wake Forest Institute for Regenerative

Medicine in North Carolina. Atala started his work on regenerative medicine and tissue engineering in the 1990s, also at Boston Children's Hospital. The interest is in part due to the mediagenic character of some of the organs that the three hundred members of Atala's lab are trying to engineer, given the importance of the "technologies of the self" (this phrase of the French philosopher Michel Foucault seems particularly relevant in this context) in contemporary societies. For example, Atala's lab has managed to fabricate lab-grown human penises, using decellularized collagen scaffolds on which cells of the patient are seeded.[28] Men who have lost their penis through genital defects, traumatic injury, or cancer surgery, or as victims of violence, and transgender people who seek to actively determine the physicality of their sexuality, are hopeful that this technology can help them to live better lives. The current technology for penis replacement was established in the 1970s and is rather crude. Penises are constructed with skin and muscle from the patient's thigh or forearm. Reproductive use is not achieved, and sexual function is implemented either using soft malleable rods, which are difficult to conceal, or inflatable bars with a saline pump implanted in the scrotum. Furthermore, the desirability of these phalloplasties may be questioned by some patients' "aesthetics of existence" (Foucault again), which can lead to further problems.

Atala has been growing penile cells in artificial scaffolds since 1992. In 2008 his lab had full success with rabbits. First, they implanted bioengineered penises into twelve penectomized male rabbits, and then they put the mended males together with female rabbits, hoping they would have sex. All tried to mate; in eight rabbits there was proof of ejaculation, and four produced offspring.[29]

Making a human penis is more complicated, however, mainly because it is bigger than a rabbit's. As with most attempts to engineer complex organs in the lab for transplantation into humans, the scaffold is not artificial; it is obtained from the organ of a donor. The tissue engineers wash a donor penis in a mild detergent that does not destroy the extracellular matrix, but removes all donor cells. After two weeks, a collagen scaffold of the penis is left, onto which they seed the patient's cultured cells—smooth muscle cells first, then endothelial cells, which line the blood vessels. Atala's team are assessing the structures for safety and functionality to seek approval from the FDA before they can move to a first in-man test. The preparation of the scaffold prior to cell culture and implantation, so that immune responses from the host patient are inhibited, is a particularly important issue to overcome.

In 2006, Atala and his team announced the first successful bioengineered organ transplant, a lab-grown bladder, which had been implanted into seven patients in 1999. Between 2005 and 2008 they announced the successful follow-up of four women given bioengineered vaginas. They also implanted the first urethra in 2004.

While building complex organs such as hearts, kidneys, or lungs remains far from the current abilities of tissue engineers, some of them are thinking of building patches that can be used to fix damaged organs. Teruo Okano[30] has managed to create vascularized heart tissue layers by growing heart cells on petri dishes that are coated with a temperature-sensitive polymer. When the temperature is lowered, the cells detach from the polymer in a compact layer. The layer contains cytokines (small signaling proteins that prompt the body to grow blood vessels), so that when the layers of cells are implanted in patients, they can grow blood vessels and prevent cell death. The method has been successfully evaluated in a clinical trial with heart disease patients,[31] and was approved by the Japanese government in 2018 for treating patients. A similar approach could be applied in the future to fabricate tissues that mimic those of the liver or the kidneys.

3-D BIOPRINTING

A new technology increasingly in use in tissue engineering labs is 3-D bioprinting, which can control the architecture of scaffold materials, chemical signals, and even the position of cells with superior precision. The technology has already been used for the generation and even transplantation of several tissues, including multilayered skin, bone, vascular grafts, tracheal splints, heart tissue, and cartilage, and it is hoped that it will help to build solid organs such as kidneys, hearts, and livers.

Linda Griffith-Cima at MIT, one of the scientists behind the "Vacanti mouse" experiment, has now invented a 3-D bioprinting process to create complex scaffolds for bone regeneration. Another successful example of a 3-D printed tissue is a tracheal replacement.[32] Biomaterials made on 3-D printers have the advantage that they are designed to factor in mechanics and degradation in response to tissue growth, so that the engineered shape can change over time. This is particularly important in infants and children. Tracheobronchomalacia (TBM) occurs when abnormal development causes a collapse of the tracheal (or windpipe) walls. Severe TBM in babies and small children may require the insertion of stents, and, in more-severe cases, tracheostomy (making a hole through the front of the neck into the windpipe). In 2015, successful implantation of an external airway splint was reported in three cases of severe TBM, using 3-D printed tissues shaped like the original airway of each child. The splint changed shape with the growth of the child, and was gradually resorbed as the child's natural airway strengthened.[33]

Tissues made by 3-D printing are very attractive to cosmetics companies. In 2015, L'Oreal announced the beginning of a collaboration with Organovo, a 3-D bioprinting company based in San Diego. L'Oreal has a well-established process for producing Episkin, its patented artificial skin made of skin cells donated by surgery patients and grown in a collagen scaffold. L'Oreal reportedly sells 50 mm–wide Episkin samples to other cosmetic and pharmacology companies. The artificial skin is used, for example, to study the effects of aging by exposing it to UV light or air, or to understand how human skin would react to proprietary cosmetic formulations. By hooking up with Organovo, L'Oreal is hoping to automate production, using the technologies of 3-D bioprinting. (See plate 10.)

Atala's lab is also putting considerable effort into developing a 3-D printing technique that allows the fabrication of solid organs and vascularization. In 2016, they reported on an integrated tissue–organ printer that can fabricate stable, human-scale tissue constructs of any shape.[34] The shape of the tissue is achieved by a computer that uses a detailed model of the anatomical shape of the organ, obtained by modern imaging techniques, to control the motions of the printer nozzles, which dispense cells to discrete locations. The printer is able to incorporate microchannels into the tissue to facilitate the diffusion of nutrients to printed cells, overcoming the diffusion limit of 100–200 micrometers for cell survival in engineered tissues. The paper reports the fabrication of mandible and calvarial (skullcap) bones, cartilage, and skeletal muscle. Atala's ultimate aim is to print a human kidney, as he explained in his TED talk in 2011. In 2019 scientists from Tel Aviv University reported a 3-D–printed heart-shaped construct that hit the mainstream media headlines. "This is the first time anyone anywhere has successfully engineered and printed an entire heart replete with cells, blood vessels, ventricles, and chambers," said Tal Dvir, the leading senior scientist. While their printed heart is an impressive demonstration that 3-D printing can be used to create living shapes, their heart-shaped tissue is not functional in any sense.[35]

Scientists are also developing ways to mimic the mechanical signals and stimuli that are so crucial for tissue development. In my lab, we are engineering magnetic nanostructures inside the scaffolds that can be targeted with magnetic fields to modulate the mechanical properties and the diffusion inside the material. Donald Ingber's lab at the Wyss Institute has developed a polymer that can react to changes in temperature.[36] When the polymer warms up to body temperature, it shrinks and squeezes the cells inside the material. The forces that the cells feel induce tooth development by triggering the genes responsible for mesenchymal stem cell differentiation.

ORGANS ON A CHIP

As it remains very difficult to grow big, vascularized, fully functional organs, some scientists are trying to make miniature organs inside chips, which, although they are not good for transplantation, might find applications as lab models for understanding organ function and disease and for toxicity testing of drug candidates. Organs-on-a-chip are expected to reduce or even eliminate drug testing in animals and humans. Scientists around the world are developing this approach to study cancer and tumors, inflammation, neurodevelopmental disorders, and other diseases, using the chips as miniature experimental models of disease and health. The chips are transparent and allow high-resolution imaging and monitoring of the cells. Models of the liver, kidney, gut, lung, fat, muscle, and the blood-brain barrier have all been made on a chip.[37] Scientists are even trying to combine several of these chips with different organ models to replicate the body's physiology. Organ-on-a-chip versions of the heart-liver-blood system[38] and of the connections between the kidney and the liver[39] are being developed to probe and model drug toxicity and disease. Ingber's lab has reportedly managed to keep multi-organ chip setups going for more than a month. In the future, they might be used to adjust different treatments before they are tried on a patient. Several companies such as Emulate, Hepregen, and HemoShear Therapeutics are all working in commercial applications for organs-on-a-chip. HemoShear has started a collaboration with Pfizer to understand the effects of drugs on blood vessels, such as inflammation or injury, using organ-on-a-chip technology.

Another application of 3-D bioprinting is organ-on-a-chip setups optimized to create thick tissues that are vascularized to allow for the flow of blood.[40] Using very advanced technology, which is able to use inks made from different materials, researchers have managed to recreate the proximal tubule of a nephron, the basic functional unit of a kidney for blood filtration. In 2016, a heart-on-a-chip device was reported that has integrated micron-size soft sensors to guide the assembly of the tissues and to provide electronic readouts of tissues' contractile forces and responses to drugs.[41]

USING BIOLOGY, PHYSICS, AND MATHEMATICS FOR ENGINEERING AND REGENERATING TISSUES

Apart from relying on artificial materials and cells to try to create tissues, scientists are trying a more fundamental approach by starting with small balls of stem cells that are treated so that they differentiate to form millimeter-size *organoids*. Organoids have been made for many tissues, including intestine, kidney, brain, and retina. Brain organoids self-organize into swirls that contain several types of brain cells, and they have been kept alive for about a year. Although, like the scaffold constructs, organoids lack a blood supply, they are considered a major technological breakthrough for the understanding of tissue renewal, stem cell functions, and tissue responses to drugs, mutation, or damage. Organoids have already been established as an essential tool in many basic biology and clinical applications.[42]

Another interesting approach is to try to give a model tissue the best possible approximation of the chemical, physical, and electrical signals that a normal tissue presents so that it can regenerate after an injury. In chapter 1, I introduced the groundbreaking work of Michael Levin, showing that electrical fields generated in the brain control the development of tadpoles. His method has been used to help adult frogs regenerate amputated legs, and even to create a two-headed flatworm. Levin plans to develop a "physiological phrase book" of mathematical models that will allow us to create electrical signaling programs to stimulate tissues to regenerate.[43]

While many challenges remain, the progress is rapid and accelerating. Modern tissue engineering is an applied field that integrates physical, mathematical, and biological sciences in the context of medicine. This practical application background is important for raising the funds and creating the multidisciplinary arena that bring rapid progress. Solving the problem of organ regeneration is spurring collaborations to understand fundamental biological problems and to generate new ways to test drugs without the need for animal or human testing.

Advancement will require better quantitative physical measurements and mathematical modeling to overcome the main challenges, such as transport of nutrients, oxygen, and wastes. Tissue engineering constructs need to achieve better real-time monitoring of all the relevant biological parameters, which include the concentrations of chemicals and proteins, the volumes of cells, the dynamic topological and mechanical properties of the tissue constructs, and the electrical signals that are so important for communication of cellular information. The rational way to proceed is to create both sensing technology that can generate large amounts of data on all the relevant parameters, and increasingly sophisticated computer models that can crunch all that data into specs for the rational design of future implants. The analysis of so much data will require machine learning algorithms (AI), as well as mechanistic models of the detailed functions of molecules and cells, including their complex chemical, electrical, and mechanical interplay in the multiscale three-dimensional architecture of tissues. Before tissue engineering becomes able to create a heart or a kidney, it will become the research arena where tissues are studied and modeled in a controlled setting. Rather than learning biology by observing it in real organisms, tissue engineering challenges us to understand biology by creating it.

The ever-regenerating body remains a distant aspiration, but on the way to achieve it, many fundamental and groundbreaking discoveries will be made. Plausibly, the progress will render organ transplantation obsolete by the end of the twenty-first century. Apart from regenerating worn-out and damaged organs, the new frontier of integrating artificial and natural materials in our bodies has the potential to enhance our senses, minds, and muscles with hitherto unavailable capabilities, leading to entirely new psychological and social scenarios for the human species.

THE FIRST BIOHYBRID, TRANSMATERIAL ROBOT

Beyond human enhancement, the creation of artificial biohybrid forms and applications will usher in an increasingly diverse "transmaterial" ecosystem.

In July 2016, Kevin Kit Parker and his collaborators introduced the first tissue-engineered soft robot (plate 11). Parker and his collaborators built a 1/10th-scale version of a stingray fish with a microfabricated gold skeleton and a silicone body powered by rat heart muscle cells.[44]

The ray is powered by 200,000 heart cells placed on top of the silicone. To organize the cells in a pattern similar to the muscles in the real ray, the silicone presents a template made from one of the proteins of the extracellular matrix, fibronectin, to guide the cells' attachment. The heart cells were genetically engineered to respond to light cues,[45] so that the undulatory movements propelling the robot through water could be controlled by light. The speed and direction of the ray was controlled by modulating light frequency and by independently stimulating the right and left fins, allowing the biohybrid machine to maneuver through an obstacle course. While the robot ray could not compute the movements itself, it is not difficult to see how that type of capability could be implemented in the future. This type of research could also be useful for learning how to build a heart that is transmaterial—half muscle, half machine.

Building the robotic ray took four years. It took a lot of work to figure out the muscle architecture of real rays, and then to design a miniature simplified disposition of the cardiac cells in the robotic ray to mimic the movement of the real one. Parker regards the stingray robot as art as well as technology: "Everyone is going to see something different in it," he says. "I'm looking at it and I'm trying to understand the heart—and impress my seven-year-old daughter."[46]

The robotic ray sums up how biological and physical sciences come together to address medical problems, but then, their experimental transmaterial offspring leaps beyond its practical origin, seizes the imagination, and pushes us into transformative new narratives and realms of invention.

We react to the vision of the robotic ray with surprise at the materialization of an ancient dream that humans have prophesied, but always from a safe distance. Suddenly, the future is here. We in awe of both the genius of these inventions and the risk of the transformative power they unleash. What are we going to do? What is it pushing us toward? Are we prepared for this?

Tissue engineering is perhaps the richest example of the dual character of our age of discovery. Convergence, creativity, and genius unleashed could solve the biggest medical problems; they could also create even bigger problems. Both the dreams and the nightmares of our species are closer than ever to coming true.

We could use the new and powerful technologies now becoming available to transform ourselves—or, Foucault warned in the 1970s, powerful institutions, states, and companies could use them for "the subjugation of bodies and the control of populations."[47] The philosophical, social, legal, political, and ethical implications of these new powers will generate furious controversies with very large consequences. The new science will have to navigate a minefield to achieve the lifesaving dreams of its practitioners—heirs to the positive spirit of Loeb and the pioneers of regenerative medicine.

5

CONCLUSIONS

Life Changes Everything

Our experience of the present times is marked by the accelerated deployment of technology, which is transforming life on Earth in all its manifestations, at every scale. In this dynamic, ambivalent scenario, full of promise and threat, life itself has become the main subject of scientific research. The knowledge accumulated by all the sciences in the previous four centuries has finally reached the capacity to interrogate, understand, modify, and utilize our own biological complexity.

The emergence of technical and theoretical tools to investigate and model life in all its intricacy is creating radical new prospects for the study of biology and the practice of medicine. While the previous generation's reductionist view, streamlining biology to molecules and genes, still prevails in much biomedical research and in the media and popular culture, a quieter wave of radical change is rising. The new quantitative biology, transformed by the intellectual framework of physics and facilitated by nanotechnology (the study and engineering of matter at nanometer scale), seeks to integrate nano-size genes and molecules within the organizational principles of the universe. The new physics of life tries to find the underlying rules that allow complex biological behavior to *emerge* across the scales, from nano (atoms and molecules) to micro (cells and their substructures) to macro (tissues, organs, and whole multicellular organisms). As I summarized in chapter 1, this way of interpreting life puts biology at the interface of mathematical, physical, and engineering sciences—and will radically change the way we find, interpret, and treat disease. More profoundly, it shifts the "scientific culture" and re-homes life and human existence (and history) within the continuum of the cosmos itself. It not only opens the way to a deeper understanding of biology and matter, but also invites us to revise our position with respect to nature—to shift the very axis of the cultural history of humanity.

Chapter 2 shows how quantitative biology has emerged hand in hand with fundamentally new materials sciences, which enlist the newly understood powers of biology to pattern matter with atomic precision and unprecedented functionality, crafting tiny structures and machines out of proteins and DNA. Learning by "making matter as biology does," scientists uncover the physical principles underlying biological architectures and the emergence of life itself at the nanometer scale. This gives us the potential to create a future of "radical material abundance,"[1] in which biological and materials sciences merge to fulfill our technological and medical needs and dreams. The new bio-nanotechnologies benefit from combining physics, biology, chemistry, machine learning, citizen scientist communities, synthetic biology, and mathematical modeling. Inspired by medicine, they are already producing astonishing breakthroughs, such as designer virus-like structures with the potential to evolve and revolutionize vaccines, or DNA robots able to assemble otherwise hard-to-synthesize chemicals.

From the point of view of chapters 1 and 2, the first wave of twenty-first–century nanomedical research (summed up in chapter 3)—which strove to improve drug delivery by loading drugs onto nanoparticles to target tumors by trial and error—already looks obsolete. By mirroring the strategies of pharmacology research, nanomedicine largely reproduced pharmacology's failures. The lessons learned reaffirm that biology's multiscale complexity cannot be sidestepped when designing new treatments. Although the early stage of biophysical knowledge and technology makes it as yet impossible to design unerring drugs or treatments, we can no longer ignore the limitations of the molecular-fundamentalist view of biology.

Drug discovery is a huge global enterprise, and it is hampered by a corresponding inertia. It is unlikely that the near-exclusive reductionistic focus on molecules will change in the near future. Aside from economic and structural constraints, the education of current biomedical scientists does not allow them to move easily to new models of disease that require a strong mathematics and physics background. This means that things will move slowly, at least in countries where the pharmaceutical giants and the academies that have evolved to feed them are entrenched. It is likely that gradual robotization of the routine tasks currently performed in biochemistry, cell biology, and pharmacology labs will contribute to the acceleration of change.

In the meantime, the reductionist vision of biology continues to spread. A very large amount of research is devoted to obtaining detailed information on the gene expression and protein and chemical content of cells, using high-throughput techniques to bring biology to the age of "big data." The so called

"-omics"—genomics, transcriptomics, metabolomics, and proteomics—are used to collect petabytes of data (a petabyte is one million gigabytes) for the construction and validation of models that do not often take into account the physical reality of the cell. By correlating genes and the epigenetic labels present on them with the proteins expressed, this approach tries to codify the output of the algorithm of life, but seems incurious about the actual physical mechanism of the algorithm, and largely ignores the principle of emergence: that in life "the whole is more than the sum of the parts." This has consequences not only for our understanding of life, but also for the success of research on drugs and drug targets.

While these "big data" approaches may yet prove a useful but costly way to identify some molecular targets, future drug design needs to take into account the whole physical picture of the function of proteins within their cellular and tissue environments. Otherwise, drugs will simply have a very hard time finding their targets, or cells will render them inactive. Bringing in more quantitative biology approaches should also narrow down the efforts of the algorithms built to search for correlations between "-omics" outputs—an amount of data that threatens to be too large to actually be useful, despite the power of new computer hardware and AI.

Machine learning and AI can be of the most use if guided by actual mechanistic hypotheses. To use a simplistic analogy, the current big-data "-omics" strategy is akin to mapping the position of every star in the universe over a long period of time, then asking a machine learning algorithm to make sense of the relationships among the movements of all of them . . . while ignoring the existence of gravity.

It is to be hoped that biomedical research communities will marshal their efforts into better strategy and coordination, which might lead them to the same conclusion that astronomers and particle physicists came to long ago: collaboration in large international research projects is the best way to tackle very complex scientific problems.

In the current landscape of drug design and delivery, it is no wonder that the most promising strategy for fighting cancer is immunotherapy. Away from the dogma of genes and molecules, and from brute force approaches to calculate the whole from its parts, immunotherapy uses biology itself to fight disease, thereby integrating all the scales necessary to do it. Its success will not only breed better treatments, but also more of the multiscale, quantitative type of research that is necessary for biomedical progress. The recent research surveyed in chapter 3 suggests that nanotechnology could help to overcome immunotherapies' current limitations and to increase the robustness of the immune system's antitumor response. As we saw, it is much easier to target nanoparticles to dendritic cells of the immune system than it is to cancer cells. Nanoparticles can also be used to produce an immune response in the local environment of the tumor while reducing systemic toxicity: by prolonging the particles' retention at the tumor site, they can release extremely potent immunostimulatory molecules and still promote a controlled response in the whole body. A particularly interesting approach is the use of implants to recruit and stimulate immune cells to attack tumor cells, working effectively as time-release cancer vaccines.

More-effective, more-targeted drugs alone, however, will still not be enough to bring about the revolution in medical research that we are striving for. We need to be able to understand and predict the effect of a drug in an individual. Without that understanding, drugs that could be useful to the majority of a patient population often cannot be approved because, for example, they produce intolerable side effects in a smaller but still sizable number of patients. Clearly, analyzing complex networks of genes and proteins alone won't be enough to design cancer therapies that work for every patient.

To advance toward personalized medicine, we will have to go beyond accumulating data on isolated cells in the lab, and become able to gather physical and biochemical information directly from molecules and cells inside living organisms, in real time. Groups around the world are already designing implantable biosensing devices, many of them nanoparticles or nanostructures, that are able to detect bacteria, viruses, or molecules—such as antibodies, other proteins, DNA fragments, or glucose—react to them, and even communicate their status to the outside world. (The first applications for diabetes, such as contact lenses that monitor glucose levels, or implantable materials that can both detect the concentration of glucose and react with a controlled release of insulin, are being tested in the lab.) Several detection strategies are being investigated, from the electrical properties of graphene or carbon nanotubes to nanomaterials that change color when the molecule of interest binds to them. Many of these systems use antibodies to home in on a particular molecule, but as we have learned in the last fifteen years, reaching specific molecules with antibodies and simultaneously being able to detect the binding to the target is not easily achieved. Extracting meaningful data from biological organisms is still a challenge, due to the size of the molecules and the complexity of the biological environment.

Both in drug delivery and in biosensing, the failures and successes of current strategies are spurring scientists to peer deeper into the basic mechanisms by which molecules interact at the nanometer scale, and the integration of those nanomechanisms at the level of cells and tissues.

Here's where big data and algorithms come in. To build new models that incorporate the hierarchical, multiscale physical nature of biology, biosensing devices will need to gather relevant real-time data from large samples of patients and healthy individuals (male *and* female; surprisingly, or maybe not, most current medical research is done on male volunteers and male rats[2]). To analyze the big data from these future biosensing platforms, it will be necessary to develop and collaborate with deep learning algorithms. The goal is to develop multiscale mechanistic models that explain biological function from the physics of the nanoscale up through the interactions that link the scales. These models, in turn, will feed back into the algorithms, refining and being refined by them.

Tissue engineering, the subject of chapter 4, is emerging not only as the field that may enable the repair and even regeneration and replacement of organs, but as an arena where fundamental progress will be made in the basic science underlying biology and medicine. Studying (or, indeed, even identifying) all the relevant quantities in a large living organism is still an impossible task, but tissue engineering can construct artificial biological tissues in which all kinds of interactions across the scales (chemical, physical, electrical, mechanical, genetic) can be studied in a controlled environment and can begin to inform model building. At the same time, experimental biosensors for constant monitoring of cell cultures and organ-on-a-chip constructs will start to generate real-time data on living tissues, enabling the development of technology that can later be used in vivo. Creating biosensing technology that monitors living processes to molecular precision and mechanistic models of tissues that link the molecular with the macroscopic will arguably be the most important contributions to medicine and biology of tissue engineering in the near future.

Tissue-engineered models are also very useful for understanding and modeling targeted drug delivery, and it can be predicted that such models of human tissues and organs will eventually replace animals in drug testing.

As the technologies progress, increasingly sophisticated and automated cultures of living tissues will replace much of what goes on in molecular biology and biochemistry labs, which currently rely on endless, selfless hours of tedious and repetitive experimentation by PhD students, technicians, and postdocs. New robotic technologies will generate biological big data that will be analyzed and classified by physics-driven machine learning algorithms to help scientists generate sophisticated models of biology. As AI becomes able to relate cause and effect, it will be increasingly employed to monitor biology with molecular resolution and multiscale awareness. The devices and models created on cell cultures and engineered tissues will eventually make their way into animal research and then human medicine, finally materializing into the technology that will be able to monitor and fix our bodies, perhaps in real time and in a personalized way. This technology may be based on hybrid transmaterial devices that use both biological and inorganic materials and principles to achieve a real-time super-enhanced immune system.

We are still far from that. But the research that began at the interface of nanotechnology and biology in the context of medicine will continue fusing with the rest of the sciences and technologies as the fourth industrial revolution progresses. It will incorporate more mathematical modeling, AI, and robotics, and it will create an increasingly accurate knowledge of both biology and materials science at the nanometer scale. Nanoscale devices will be used to learn biology; biology will be used to inspire and refine transmaterial devices that mimic some characteristics of biology and incorporate some features that are not available in the biological world. The physics being applied to both fields will finally bring about a new kind of materials science that reunites biological and inorganic matter, as Kevin Kit Parker's pioneering "soft robot" that you met in chapter 4 prophesies.

Beyond medicine, far-reaching consequences for our identity and relationship with nature ensue from the incorporation of physics into our understanding of biology. Physics frees us from the one-dimensional, reductionist molecular tyranny of the genes and allows us to feel ourselves as emerging from the deep, multidimensional fabric of the universe. Ultimately, biology materializes from the interplay of forces, energy, space, and time that has given us our world, senses, intelligence, conscience—and our immense powers to learn how to heal, regenerate, and reinvent ourselves, our cultures, and our environments using technology. The new physics of life allows us to inquire more profoundly into the physical underpinning of our human intuition to do science, to dive to the foundations of inanimate and animate matter; it hints at the underlying structure that links creativity, intuition, and learning—in other words, the arts, the humanities, and the sciences.

The anticipated advent of biology to physics and engineering is relocating our point of view and producing new convergences. The questions arising from scientific research increasingly coincide with the queries of philosophy, the arts, and the humanities regarding the very nature of human existence: What is life, and intelligence, and where do they come from? What is our position in the world as we gain greater and greater control over matter and biology? To start with, these inquiries are being formulated in a purely scientific context. But as we re-pose these questions, we reconnect with and test the beliefs, intuitions, and inquiries that have preoccupied human culture since the birth of early civilizations more than ten thousand years ago. This junction of modern sciences and cultural history brings us to the crossroads where we now stand: between a path to unprecedented vistas of human advancement, and a path to the mindless exploitation of our own nature.

During the last four centuries, science led us to the technologies that we used to indulge in the illusion of complete power over nature. Current societies largely regard nature as a commodity, at the service of systems that direct technology to extract its economic value. The scientific developments that I have summarized in this book embed human biology and life itself into the substrate of this commodification of nature. Our bodies and minds are the next frontier for economic exploitation.

If we are to mature as a technological species and indeed survive as humanity, we need to confront our adolescent use of technology, and grow into new social and economic systems that empower us to deepen the meaning of our existence. At this moment in history, we have no other choice than to mature to become "part of the whole" in a physical, economic, and social sense, and to advance into the construction of a new relationship with nature that allows our survival. In the epilogue I argue that the reconnection of science and culture happening in our labs is pointing in the right direction.

EPILOGUE

BIOLOGY BECOMES PHYSICS

Our Coming of Age as a Technological Species?

The consequences of the fourth industrial revolution have been the subject of animated debate in the last couple of years; academics and authors have produced books, courses, videos, and essays in which they explore the futures that the expected revolution might bring. While some of their efforts seek a balance, there are also overoptimistic predictions of paradise. Other prophets warn of Huxleyan dystopias in which subjugated citizens will voluntarily give up their freedoms and their political and social powers to AI algorithms for a universal basic wage. They foresee a post-work society whose robots have taken over all the jobs that gave meaning to our lives. They take us to a world where a new elite of super-enhanced humans have used bio- and nanotechnologies to achieve an incontestable control over the future of humanity, and of life itself.

In this book, I have briefly presented some of the ways in which biological research has changed in the last decades. These changes are not only transforming medicine, but, perhaps more importantly, producing a profound scientific culture shift. The convergence of sciences on biology is a response to major obstacles insurmountable under current scientific paradigms, and to the need to speed up technological progress. But the incorporation of biology into the realm of physics also repositions our point of view as humans, and forces us to confront the meaning of our existence and our relationship with the rules that govern the universe itself. Some of the most important consequences of these changes have not been dealt with in most of the public debate about technology, mainly because scientists have not been very active in thinking about them.

The emergence of tools that allow a mathematical description of biological phenomena makes it possible to develop new engineering powers, which could lead to a fundamentally new scenario. On the positive side, it will enable unprecedented technological and medical advances. On the negative side, it may lead to the commodification of our own biology and to an even deeper gulf between the haves and the have-nots. In this situation, as scientists and as humans, we yearn for new cultural narratives that will bring new depth and meaning to our relationship with nature, and will guide our use of technology toward a future of human advancement and improvement rather than a high-tech apocalypse.

In this epilogue, I want to show some ways in which scientists are joining the effort to create and decipher this incipient new world by interacting in new ways with society as they strive for a new sense of scientific purpose.

SCIENTISTS STRIVE FOR NEW TECHNOLOGICAL CULTURES

The changes wrought on societies by globalization and the deployment of increasingly transformative technology have also altered the way scientists do science, and stirred them to engage with their work and its social context in more-collaborative ways. Three trends show how scientists are contributing to the creation of new technological cultures that will keep the sciences conscious and careful of their impact on the way we live.

First, scientists are taking the lead in anticipating the risks of the technologies they develop. In fact, nanotechnologists, as a community, were the first to embrace this commitment, and they have been at the forefront of the development of the "Responsible Research and Innovation" (RRI) agenda for the last decades. The powerful potential of their science and technology has made nanotechnologists very conscious of the potential for good, but also the potential for misuse and mismanagement. Nanotechnologists pioneered technological sustainable development and the inclusion of responsible principles in the scientific process. They did so by engaging with social scientists, and by lobbying institutions such as the European Union for the funds to work out the practicalities of implementing RRI in the scientific day-to-day. The first RRI program started in the late 1990s. RRI principles are currently integrated into EU-funded research projects, emphasizing five aspects: public engagement; distribution of research outputs (online, free of cost or other barriers, currently known as "open access"); gender equality (both in the composition of research teams and study populations, and in the potential impact of the work); ethics; and science education. Rather than just waiting for regulators to ban a product once the damage has already been done, nanotechnologists left the ivory tower and have been vigorous in engaging with the public, politicians, and regulators to discuss the potential risks (e.g., toxicity and ecotoxicity) along with the possibilities. NGOs such as Matter[1] have led the way in creating partnerships between academia, companies, international organizations, and the wider society to direct technology to create more-sustainable, socially responsive innovations that benefit us all. Centers, programs, regulatory frameworks, and grants, very much driven by responsible scientists, have been established worldwide to ensure that the new technologies integrate the sciences, regulation, sustainable development, and governance, including issues of equality. Much of this work started in nanotechnology centers, and many lessons about self-regulation can be learned from nanotechnologists' efforts of by other fields, such as the emerging artificial intelligence community.[2]

Beyond work in academic settings, tech workers are finding it increasingly difficult to participate in the development of technologies that can be used against democracy, the general public, or human rights. In the last year the mainstream media have started to feature news about the Tech Workers Coalition and their successful activism in pushing U.S. tech companies to listen to their workers in cases where the uses of technology raise serious ethical issues. Tech worker activism forced Google to abandon Project Maven, which provided surveillance technology to the U.S. Department of Defense, in 2018. Tech workers are currently leading a campaign against Microsoft's work with Immigration and Customs Enforcement (ICE), especially after the shocking images became public of the separation of children from their parents at the U.S.–Mexican border.

A second trend is the surge of citizen science projects, such as Rosetta@home and Foldit, that I discussed in chapter 2. Together with initiatives in other fields, such as GalaxyZoo, they are connecting academic science with hundreds of thousands of participants all over the world at a level that has never been achieved before. In citizen scientist projects, the public not only contributes time and processing power on their computers to amass the power necessary to calculate very difficult problems; they are also encouraged to participate in the actual process of solving the scientific questions, to learn about the science and engage with academics, and even to become coauthors of scientific publications.

The surge of groups that get together to perform experiments in amateur science labs is also growing. The society DIYbio.org was founded in 2008 to create "a vibrant, productive and safe community of do-it-yourself biologists." Central to its mission is the belief that biotechnology and greater public understanding of it have the potential to benefit everyone. DIYbio.org has already established more than 100 local groups all over the world, which hold regular courses, workshops, and meetings. The company Amino Labs offers kits for amateur scientists to learn the basics of synthetic biology, including how to extract DNA and how to genetically modify bacteria to produce proteins. The era of the citizen scientist has started.

Perhaps the most potentially transformative of all these trends is the erosion of the boundaries between the arts and technology and science, which are being breached from both sides. Artists collaborate with scientists and technologists. Artists become scientists, scientists become artists. This modern combination of the sciences with the arts generates works that increasingly leave the art galleries and the research institutions to seek dialogues with the public in unusual places: streets, hospitals, shopping centers, pubs. Examples can be found everywhere; they sprout and spread fast, building on scientific public-engagement activities such as Soapbox Science (featuring women scientists) and Pint of Science, and they often intermingle with the activities of citizen scientist networks. The general public are eager to know more and to be inspired by the science, regardless of political inclinations and perceptions of social status, and they naturally recognize the alliance of the sciences and the arts. Very often these activities are led by women, timidly yet ambitiously hoping to grow into a stealthy transformative movement. Often, in this and other contexts, in both the developing and developed worlds, women are putting their hopes in science and technology as a path to a fairer and more meaningful society.

Artists and scientists are responding to the need for new spaces for creation and reflection in the face of the radical changes new technologies bring to our bodies and health. An example is artist Sofie Layton in the UK, who develops art to help patients and doctors understand the new technologies appearing in hospitals. By merging science, art, patient care, and medicine, she created the project Under the Microscope at the Great Ormond Street Hospital in London in 2016. Through workshops with clinicians and researchers, in which artworks were created, her project explored how children and their families understand diseases and the complex modern treatments they are subjected to. Bringing the new medical sciences to patients is complex, and there is a need for new ways to both understand and shape the uses of the new technologies. Mark Ackerley created DNA Melody while he was working for 23andMe, one of the biotech companies that help customers research their ancestry by analyzing DNA from their saliva. Ackerley turned DNA snippets into a musical score, using rhythm, pitch, timbre, and key to translate genes into melody. Better known are, for example, the works of the Icelandic artist Björk, especially her 2011 album *Biophilia*, in which she combined science, digital technology, and traditional instruments to create songs and apps about physics and biology, crystals, atoms, viruses, and dark matter.

These works seek to prepare us, as well as to prepare the technologies to serve us, not only in an economics-centric way, but to foster positive cultural advancement. They seek to explore the changes that technology causes to our identity, our health, our biology, our relation with the world, and our perceptions of reality. This activity reflects our natural need for art as a means to understand and to create the narratives, metaphors, and cultures that will enable us to take constructive, positive control of our destiny and our relationship with nature. We need art to link the modern world with our traditions, cultures, and myths; to build our space in history, and to create a collective sense of the future. Art is in fact the best way to take the reins of technology, to shape it, to imagine it, to give it meaning within our shared values of humanity and history, to make science merge with our ever-evolving identities, and to anticipate and mitigate its threats.

There is an underlying theme to all these trends: the drive of scientists to democratize science, to create platforms and frameworks of collaboration with the public, and together to imagine better, more diverse, and more equal futures for us all.

TECHNOLOGY AND EQUALITY

A main preoccupation that drives scientists to engage more profoundly with society is the effect of technology in an increasingly unequal world—and the cultural perception of technology as a source of inequality. As with most sources of economic profit, the rewards and benefits of science and technology are lopsided in our society. Much of the Western narrative of technology is about the surprise it produces, the shock in our lives, and the fear of joblessness and human redundancy. This narrative stems from the fact that technology has mainly been used to control and exploit nature. Now, we unsurprisingly foresee that this technology will inevitably be used to render the weak (and the not so weak) socially and economically unnecessary, or even worse, make them fodder for the dystopian exploitation of human biology itself. Science and technology promise to improve our lives, but also to remove the last of the little control most people feel they have over technology's use and exploitation.

The possibilities that science and technology are bringing to us will probably make the twenty-first century the best and most exciting time to be alive—but only for the very few who can benefit from them: the educated, the well-connected, the powerful, and the rich. Inequalities in income and in access to health care and education are seriously threatening to derail the fabulous possibilities of our time. The future of "radical abundance"[3] that scientists in labs all over the world increasingly feel to be within human reach will not be brought about if the benefits of technology are not shared more evenly.

Yet technology is not an external force acting on society. Technology's applications arise from the conditions set and decisions made by scientists, technologists, the funders of research, the regulators, the workers, the consumers, and, ultimately, the exploiters and owners of the means of production. Society can determine the use and fair distribution of technology's bounty. Robots will produce unemployment if the main purpose of their deployment is productivity and revenue for their owners, but this is not the only possible outcome. They can also be used to make our lives more comfortable and fairer. As I have briefly shown in a few examples, scientists are increasingly active in engaging with society to create not only economic revenue but also social and cultural value. Technological change can and should be modulated by the dialogue between what is possible and what is desirable, and scientists should not be written out of this dialogue.

I would argue that technology and equality can and should feed each other. We need policy creativity for anticipatory, adaptive governance to ensure that science and technology are used to reduce inequality, rather than to become a new source of it. Such governance will, correspondingly, need science and technology to become a reality.

From my point of view as a woman, mother, physicist, and educator, the vision is clear: the potential is huge. In the lab, the international and multidisciplinary tenor of our research at the interface of nanotechnology and biology empowers our female and male students from all backgrounds, enhancing their scientific and technological creativity and their social and industrial entrepreneurship. Many applications of the new materials sciences emerging at the interface of biology and nanotechnology are potentially cheap and easy to implement, requiring minimum lab infrastructure. With the right framework, the new technologies should become global forces to reduce national and global inequalities. We should embrace that potential. Nanotechnologists are already endeavoring to democratize scientific tools to produce cheap, easy technology that can get to people around the world, such as biosensing devices on paper strips. They join others in "frugal design technology": for instance, the Raspberry Pi Foundation, born from the University of Cambridge in 2009, created the Raspberry Pi computer, which costs around $35 and has sold over 10 million. In 2017 the press brought us images of the 20-cent Paperfuge, a centrifuge made of paper by Stanford engineers, able to separate blood into its components using the principle of the whirligig toy. Another example is the Foldscope, a paper microscope that costs less than a dollar.

Becoming better at controlling matter naturally brings out the human instinct to produce technology that is cheaper and more democratic. Contrary to most of the commentary that we are subjected to by academics and the press, technology in itself naturally promotes equality by making production better, cheaper, and more sustainable, and by inspiring scientists to pursue simplicity and practicality. It takes a conscious political and economic effort to create and maintain the structures that generate inequality out of technology, and not the other way around.

The seeds of the technology's potential to set off a global wave of successful, transformative entrepreneurship have already been planted. Students are attracted by the opening to disrupt economic systems that do not offer them promising futures. They want to create technology that changes their world for the better, and this is not only happening in Boston, Silicon Valley, or Oxford. Technology can be a very practical solution for many local problems, and not only in the developed world; many in developing countries are putting hopes in it.

The convergence of the sciences around biology offers great development opportunities. For example, most Asian countries, which don't have strong pharmaceutical industries (apart from Japan), appreciate the possibility of developing medical technology that can be disruptive of the current status quo. They foresee an opportunity for growth and even global dominance in technologies that will shape the future; this is clearly reflected in the research budgets for physics, engineering, and materials science projects touching on biology and medicine in, for instance, South Korea, China, Singapore, and Taiwan. It is clear that the research

I have presented in this book is starting to influence global economics and geopolitical strategies.

Doesn't the developed Western world's fear of technology and largely negative vision of the future also reflect the fear of the rich and powerful that they will lose their privileged position in the world, and even Western societies' fear of falling from their cultural and economic dominance? Isn't it a kind of ambivalent game where the people who feel entitled to produce and exploit the technology also produce the fears as a means to facilitate its control, as well as to prevent its misuse? Does this ambivalent position reinforce the current trend in most of the Western world to reduce budgets for education, basic science research, and collaboration that may threaten the dominance of some of the main industrial players in future technologies?

While it is surely a good idea to interrogate and regulate technologies such as AI, robotics, biology, and nanotechnology, it is also true that for the large corporations that dominate markets, many of the products and applications of the new technologies are disruptive, threaten their current models of economic sustainability and growth, and are being developed in places outside their traditional control. These corporations have the power to put the brakes on research and development that threaten their control by effectively lobbying governments. The media and the entertainment industry can divert attention from the actual power struggles by creating narratives that contribute to the public's feelings of alienation and frustration and turn them against an elitist class of scientists, technologists, and experts. Fear of technology is used as a political and economic weapon as powerful as technology itself.

CREATING VISIONS OF POSITIVE TECHNOLOGICAL FUTURES

And so, the growing fear of a technology that might be soon have the power to derail human life on Earth coexists with the scattered attempts of scientists, artists, and citizens to imagine how that same power might be harnessed to create a better world—one where life is possible, fairer, and more meaningful.

In this complex scenario, our disoriented hopes, fears, and efforts are struggling to birth new cultures that make our survival possible. How do we construct narratives and visions of the positive futures that we would like to inhabit? How do we figure out what we really want so that we can work toward its realization?

The fear of technology in the West has deep historical roots. The harshness of the first industrial revolution in England left deep scars on the psyche of much of the West in relation to technology, as best summarized by Mary Shelley in *Frankenstein; or, the Modern Prometheus* (published in 1818, when she was twenty) and followed up by the works of George Orwell and Aldous Huxley. This narrative born of the Anglocentric experience is still prevalent—in the new works of academics forecasting the future, and in the myriad cultural representations of dystopian futures.

Other Western countries have had very different experiences with technology. In my native Spain, technology is loved by old and young; a solar panel on the roof is displayed with pride by the owner, regardless of social origin, as a sign of liberating modernity, while to my amazement I have often heard in rural settings in England that solar panels on roofs look "horribly modern" and spoil the "country living" experience. This is not just for aesthetic reasons, but out of technophobia. The dominance of English as the global language for culture and academic writing has spread a particular vision of technology that evolved from the foundations of modern Western capitalism. That negative view resonates again now with the meme of "precarity" and the scarcity mentality of much of the existing job market. The problem with this negative forecasting is that it leaves us unprepared for progress; without a vision of what we want, we won't be able to shape the future in any way.

The relationship between society and technology is very different in countries where science is seen as a source of well-being. A useful example to look at is Japan. After the Second World War, Japan skillfully navigated complex geopolitics to become Earth's second largest economy—thanks to a growth strategy based on technological development, with negligible unemployment and an unprecedentedly fair distribution of wealth among—mostly male—workers and their families. Technology is felt as a force for good not only in Japan, but widely across East Asia. Much of modern popular culture there puts its faith in a future in which the harmonious integration of humans, nature, and technology becomes possible. Japanese fiction has fewer Frankensteins.[4]

Although contemporary Japan faces serious economic, social, and intellectual stagnation, fueled by fear of immigration, gender inequality, and exceptionally low birth rates, its citizens' hopes for positive transformation remain intertwined with technology. I would like to share with the reader an example from Japan of progressive, creative forces identifying ways to generate positive cultures using technology. It is a model that may inspire us to formulate our own wishes and dreams of the futures we want for ourselves.

Toshiyuki Inoko created teamLab Inc. in 2001, when he graduated from the Department of Mathematical Engineering and Information Physics of the University of Tokyo. He was interested in "technology changing the world and art changing people's minds . . . and values,"[5] so he committed himself to create digital art that would "suggest an alternative way to live in contemporary society." His dream was to live and create collectively with his friends, and so he created teamLab. Today, teamLab has become an association of several hundred "ultra-technologists," including programmers, engineers, mathematicians, architects, computer graphics animators, and other specialists, who work together in a combined commercial and artistic operation from their offices in central Tokyo.

The turning point came in 2011, with the exhibition "Life!" at the Kakai Kiki gallery in Taipei, where teamLab was invited by the artist Takashi Murakami. Since then, they are gradually becoming a global phenomenon. teamLab fuses art, technology, and the natural world with ancient Japanese narratives and artistic perceptions. They dazzle viewers with spectacular large-scale installations, immersive and interactive, that confront conventional thinking while rooting the intellectual aspects of the work in Japanese classic arts and traditions. "I want to be with people who want to take a step into the new world," Inoko says. "People who are creative, who want to change the world, these people I hope to inspire and have influence on."[6] The last two years have been a frantically productive time for teamLab, with acclaimed exhibits all over the world. I enthusiastically invite the reader to visit their website and, if opportunity arises, to pop into one of their exhibitions and join their visitor-participants in co-creating the experience.

The works of teamLab respond to the disconnected and contradictory feelings that technology evokes—the mixture of anticipation and dread, the sense of doom created by our histories and our social and economic conditions intertwined with the certainty that we can create beauty and happiness if we put the right pieces into place. In our disoriented present, teamLab's fusion of artistic vision and cutting-edge technology shows a beauty that invites us to embrace our coming of age as technological species.

"WALK FORWARD IN THE RADIANCE OF THE PAST"

To transmit the inspiration and hope I take from teamLab, I have chosen one of their works to conclude this book, because it mirrors and sums up much of its content: teamLab's *Enso*. *Enso* means "circle" in Japanese. Drawing circles in one brushstroke has been a central activity of Zen Buddhist calligraphers since the thirteenth century. Each *enso* is different; it reflects the body and soul of the calligrapher, but also the hearts and minds of those who view it in that particular instant. It links space, time, human perception, technology, and action; it represents enlightenment, truth, the entirety of the universe, and human complexity and equality. From a biological physicist's point of view, *enso* is a kind of two-dimensional projection of the state of our consciousness and our biology in an instant, always different, always changing. A complex convolution of all the signals that we are striving to suss out with our nanotechnology-based biosensors, a representation of each of the mathematical models of health and disease that we will build with our future technologies. (See plate 12.)

teamLab has succeeded in reinventing the *enso* in three dimensions, using "Spatial Calligraphy." With their technology, one can paint a three-dimensional circle suspended in space. The algorithm lets the *enso* evolve in time, using accurate computer models of the fluid dynamics of the virtual ink as it merges with the space around it. Inputting wisdom, traditional knowledge, and cultural history into algorithms, teamLab is working in parallel to (but independently of) the fusion of modern medicine, nanotechnology, computer science, and physics, creating a deeper relationship with nature by accessing the depths of our own biology using technology.

In this book I have observed that the incorporation of biology into the realms of physics and engineering forces us to feel ourselves as emerging from the very rules that govern the universe. Knowing how those rules govern life will unlock unprecedented power, but will also force us to merge with nature in a more profound way. Failing to do so will lead to technological stagnation and possibly to our extinction. We are both susceptible to being commodified by our immature use of technology and capable of unleashing a radiant future of human flourishing if we can create cultures that nurture our technological maturity. As the young Swedish climate activist, Greta Thunberg, and so many schoolchildren around the world so wholeheartedly remind us, we contemplate the end of our technological adolescence confronting dismal prospects, but recognizing within ourselves our life path to adulthood.

The new science addresses, for the first time in human history, deep questions such as the origin of intelligence and of life itself, using the tools of physics. By doing so—mirroring teamLab's *Enso*—it closes a cultural loop of thousands of years, reconnecting with early civilizations' powerful intuitive attempts to understand our nature and our position in the universe. Naturally, in the twenty-first century, the sciences and technologies converge with the arts and humanities, as we all struggle to find a fairer and more democratic way out of the disillusionment, the angry cries of nationalism, anti-intellectualism, sexism, and xenophobia that characterize so much of our world today.

We are at a crossroads. We can either heed the loud call of the fearful, reductionist pessimists, and become smaller, or we can seize the unique opportunity for humanity and emerge into an improved technological, social, and economic democracy. By gathering our collective narratives, we can dare to "walk forward in the radiance of the past."[7]

Notes

INTRODUCTION. SCIENCES CONVERGE INTO BIOLOGY TO TRANSFORM HEALTH

1. Although most eukaryotic cells are 10–100 microns in size, some can be very large. For example, an ostrich egg, which has a diameter of 12 cm, is a single cell; the dorsal root ganglion, a cell carrying sensory information from the body to the brain, can reach 2 meters in length in a tall human and 25 meters in a blue whale; the laryngeal nerve of a giraffe is a single cell that can extend as far as 3 meters; some giant amoebas are single-cell organisms that can measure 40 mm.
2. In this book I refer to "intelligence" in a very broad sense, as the capacity of a living organism to make sense of its environment in order to survive.
3. Biological information involves extremely large quantities and multiple types of data, and hence belongs to the category of "big data"; other kinds of big data include the information collected through the internet from all its users, or the data collected in CERN and other particle accelerators and by the Hubble and other deep-space telescopes.
4. Quantum mechanics is the branch of physics that studies the behavior of very small particles, such as atoms and subatomic particles—electrons, neutrons, and so on. Early twentieth-century experiments gradually demonstrated that the "classical mechanics" developed by Newton and others to describe macroscopic forces and movements of objects could not be applied to atoms and smaller particles. Quantum mechanics differs from classical mechanics in that in small systems (e.g., atoms), quantities such as energy and momentum are restricted to discrete values (they are *quantized*); quantum objects behave as both particles and waves at the same time (*wave-particle duality*); and there are limits to the accuracy with which quantities can be measured (this is the *uncertainty principle*). Spooky characteristics

of very small units of matter, such as *quantum entanglement*—in which two or more particles' properties are dependent on each other regardless of how far apart they are in the universe—are currently the subject of much investigation for applications such as quantum computers.

5. In crystalline solids, atoms are arranged in perfectly ordered 3-dimensional arrays.
6. For reasons that resonate with much of the creative and intellectual community, science has been imbued with a sense of altruism that facilitates scientific progress—but, on the other side, the professional "precarity" increasingly suffered by workers in research and technology is exploited to pursue goals that do not benefit society. This is a complex topic which is outside the scope of this book.

CHAPTER 1. EMBRACING BIOLOGY'S COMPLEXITY, AT LAST

1. With great intuition, the founder of microscopic anatomy, Marcello Malpighi, predicted in the seventeenth century the existence of *machinery* behind biological complexity. "The operative industry of Nature is so prolific that machines will be eventually found not only unknown to us but also unimaginable by our mind," he wrote in *De Viscerum Structura*, in 1666.
2. As the publisher summed up Denis Noble's view in his celebrated book *The Music of Life: Biology beyond the Genome* (Oxford: Oxford University Press, 2006).
3. The reader may recognize the Eames name from the couple's work as designers of, among other things, the famous Eames chair.
4. There is no consensus about the conditions that led to the emergence of prebiological molecules. Some argue that life emerged in the hot water deep underground or near hydrothermal vents in the ocean; others, that life emerged from cold salty water trapped in ice. It is also possible that life arrived from other planets, carried along by a meteorite, or even that life could have sparked from a meteorite collision.
5. Jeremy England, and many others since Erwin Schrödinger wrote *What is Life?* in 1944, have argued that the formation of complex molecules and evolution itself are programmed into the physics of the universe, in processes that are not in thermodynamic equilibrium—that is, where energy is dissipated into the environment, e.g., as heat. This use of energy enables the formation and evolution of *complexity*, the reduction of entropy that is char-

acteristic of life. See, for example, a summary of England's thinking in Natalie Wolchover, "A New Physics Theory of Life," *Quanta Magazine*, January 23, 2014. My own current work seeks to apply these ideas to the study of growth and shape in biology.

6. Others have written about reductionism using the images from the film *Powers of Ten*—for example, Robbert Dijkgraaf, "To Solve the Biggest Mystery in Physics, Join Two Kinds of Law," *Quanta Magazine*, September 7, 2017. In fact, *Powers of Ten* has inspired much writing, filmmaking, and philosophical writing. Another recent example is Derek Woods, "Epistemic Things in Charles and Ray Eames's *Powers of Ten*," in *Scale in Literature and Culture*, ed. Michael Tavel Clarke and David Wittenberg, 61–92 (Cham, Switzerland: Palgrave Macmillan, 2017).
7. Microscopes able to see cells were invented in the seventeenth century by Anton van Leeuwenhoek, a Dutch scientist and tradesman. He developed a technique for making extremely curved glass to dramatically increase the resolution of optical devices. His technique may have been inspired by the lens-grinding techniques of Baruch Spinoza, the philosopher (and craftsman) whose writings later inspired Einstein's views on the origin of the universe (Spinoza lived a few miles from van Leeuwenhoek, and they had friends in common). Van Leeuwenhoek's initial purpose was just to be able to count the number of threads in fabric to determine the quality of woven goods. But his microscope was much better than any other constructed before, and this allowed him to be the first human to see individual microscopic organisms, such as bacteria and yeast. The secrecy with which he protected his method of constructing lenses is as renowned as his discoveries.
8. Robert Hooke in England managed to construct a microscope that allowed him to observe that plants were made of cells, a term he coined to describe the plants' building blocks because their packing reminded him of the small rooms where monks used to live in in monasteries.
9. Certain aspects of Brillat-Savarin's "Gastronomie Transcendante" inspired the studies of the nineteenth-century Dutch chemist Gerardus Johannes Mulder in his search for the "most essential substances of the animal kingdom." From his studies dissolving silk, beef gelatin, egg whites' albumin, and other animal and plant substances with acids and caustic potash, he inferred that there was a fundamental unit that was present in all biological substances. His Swedish colleague Jöns Jacob Berzelius suggested that he could call these units *proteins*, from the Greek *proteios*, meaning "of first

rank or position." And so proteins were officially discovered and named in 1839. Mulder immediately made the link with the nutritional value of certain foods and suggested that it would probably be very useful to study proteins further in the future. The early history of proteins was written by chemists who developed an intense study and rapid understanding of the chemistry of life.

10. As I write this sentence (in January of 2019), Cold Spring Harbor Laboratory has announced that it is revoking all titles and honors conferred on James Watson (who led the lab for many years) over "reprehensible" comments connecting DNA, race, and intelligence. DNA and politics are almost always intertwined in not-very-nice ways.
11. Information does not flow in this sequence in all cases, though; for example, there are also processes in which information is transferred from RNA to DNA (e.g., in retroviruses, such as HIV). Some viruses are able to transfer information from RNA to RNA to produce proteins without the involvement of DNA.
12. Stephen Jay Gould, "Humbled by the Genome's Mysteries," *New York Times*, February 19, 2001.
13. John R. Shaffer et al. "Multiethnic GWAS Reveals Polygenic Architecture of Earlobe Attachment," *American Journal of Human Genetics* 101 (2017): 913–24.
14. Evan A. Boyle, Yang I. Li, and Jonathan K. Pritchard, "An Expanded View of Complex Traits: From Polygenic to Omnigenic," *Cell* 169 (2017): 1177–86.
15. Veronique Greenwood, "Theory Suggests that All Genes Affect Every Complex Trait," *Quanta Magazine*, June 20, 2018.
16. Kat McGowan, "I Contain Multitudes," *Quanta Magazine*, August 21, 2014.
17. Elena Kuzmin et al., "Systematic Analysis of Complex Genetic Interactions," *Science* 360 (2018): eaao1729.
18. Summary of Denis Noble's theme in *The Music of Life*.
19. Noriyuki Kodera and Toshio Ando, "The Path to Visualization of Walking Myosin V by High-Speed Atomic Force Microscopy," *Biophysical Reviews* 6 (2014): 237–60.
20. N. Kodera et al., "Video Imaging of Walking Myosin V by High-Speed Atomic Force Microscopy," *Nature* 468 (2010): 72–76.
21. Adam Curtis and Chris Wilkinson, "Topographical Control of Cells," *Biomaterials* 18 (1997): 1573–83.
22. Adam Curtis, "Small Is Beautiful but Smaller Is the Aim: Review of a Life of Research," *European Cells and Materials* 8 (2004): 27–36.

23. Roger Oria et al., "Force Loading Explains Spatial Sensing of Ligands by Cells," *Nature* 552 (2017): 219–24.
24. Adam J. Engler et al., "Matrix Elasticity Directs Stem Cell Lineage Specification," *Cell* 126 (2006): 677–89.
25. Readers who have had the experience of eating the brains of animals, as we still do in my native Spain, will know what 3 kiloPascals are, although cooked brains have higher *elastic modulus* (resistance to deformation) than raw ones. In fact, once I saw a presentation by a neurosurgeon showing that a raw brain can be "poured" into a bottle or other container, and flows almost like a dense liquid, taking the shape of the container almost immediately.
26. Dennis E. Discher, David J. Mooney, and Peter W. Zandstra, "Growth Factors, Matrices, and Forces Combine and Control Stem Cells," *Science* 324 (2009):1673–77.
27. Ning Wang, Jessica D. Tytell, and Donald E. Ingber, "Mechanotransduction at a Distance: Mechanically Coupling the Extracellular Matrix with the Nucleus," *Nature Reviews Molecular Cell Biology* 10 (2009): 75–82.
28. *Tensegrity* is a term coined by Buckminster Fuller, architect, engineer, artist, and many other things, to describe his vision of a new kind of architecture, conceived as emerging from the rules of nature. Fuller created tensegrity structures that maintained their stability, or integrity, through tensional force.
29. Arash Tajik et al., "Transcription Upregulation via Force-Induced Direct Stretching of Chromatin," *Nature Materials* 15 (2016): 1287–96.
30. Arvind Raman et al., "Mapping Nanomechanical Properties of Live Cells Using Multi-Harmonic Atomic Force Microscopy," *Nature Nanotechnology* 6 (2011): 809–14.
31. The Nobel Prize in Physiology or Medicine for 1963 was awarded jointly to Sir John Carew Eccles, Alan Lloyd Hodgkin, and Andrew Fielding Huxley "for their discoveries concerning the ionic mechanisms involved in excitation and inhibition in the peripheral and central portions of the nerve cell membrane."
32. Henry W. Lin, Max Tegmark, and David Rolnick, "Why Does Deep and Cheap Learning Work So Well?," *Journal of Statistical Physics* 168 (2017): 1223–47.
33. Jim Gimzewski, co-leader of a neuromorphic electronics project at UCLA, in Andreas von Bubnoff, "A Brain Built from Atomic Switches Can Learn," *Quanta Magazine*, September 20, 2017.

34. Adam Stieg, in von Bubnoff, "Brain Built from Atomic Switches."
35. Liping Zhu et al., "Remarkable Problem-Solving Ability of Unicellular Amoeboid Organism and Its Mechanism," *Royal Society Open Science* 5 (2018): 180396.
36. "Natural Computing refers to computational processes observed in nature, and human-designed computing inspired by nature." Definition from the journal *Natural Computing*, Springer, https://link.springer.com/journal/11047.

CHAPTER 2. LEARNING BY MAKING: DNA AND PROTEIN NANOTECHNOLOGY

1. Although small proteins fold spontaneously into their correct, functional conformation, large proteins are usually corralled into small spaces created by *chaperonin* proteins to optimize the folding process and avoid misfolding errors. Some proteins don't have a single folded shape; in fact, there are many "disordered" proteins whose roles we are just starting to understand.
2. The Holliday junction is named after Robin Holliday, who proposed its existence in 1964.
3. Escher's half-brother was a crystallographer and an influence on his work.
4. Steven Poole, "The Impossible World of MC Escher," *Guardian*, June 20, 2015.
5. In fact, knotting of DNA is a problem for living organisms. The tight packing of the 2 meters of DNA in the 5 microns of the cell nucleus means that DNA is prone to tangle and knot. Evolution has solved this problem with the *topoisomerases*, discovered in 1971 and studied by James Wang at Harvard for about 28 years. Topoisomerases are enzymes that pass DNA strands or double helices through one another. In their presence, linked DNA rings or loops can come apart, and different topological forms of DNA can change into one another. Without topoisomerases life is not possible.
6. The idea of doing this kind of computation is "to use the large numbers of molecules present in a solution to perform many operations in parallel," testing what interactions are possible and which aren't. One can design DNA tiles so that their assembly can imitate the operation of a Turing machine and perform operations such as addition. http://seemanlab4.chem.nyu.edu/XOR.html.
7. Paul W. K. Rothemund, "Folding DNA to Create Nanoscale Shapes and Patterns," *Nature* 440, no. 7082 (March 16, 2006): 297–302.

8. Ebbe S. Andersen et.al., "Self-Assembly of a Nanoscale DNA Box with a Controllable Lid," *Nature* 459 (2009): 73–76.
9. Chikara Dohno et al., "Amphiphilic DNA Tiles for Controlled Insertion and 2D assembly on fluid lipid membranes; the effect on mechanical properties," *Nanoscale* 9 (2017): 3051–58.
10. Shawn M. Douglas, Ido Bachelet, and George M. Church, "A Logic-Gated Nanorobot for Targeted Transport of Molecular Payloads," *Science* 335 (2012): 831–34.
11. Sonali Saha et al., "A pH-Independent DNA Nanodevice for Quantifying Chloride Transport in Organelles of Living Cells," *Nature Nanotechnology* 10 (2015): 645–51.
12. Enzo Kopperger et al., "A Self-Assembled Nanoscale Robotic Arm Controlled by Electric Fields," *Science* 359 (2018): 296–301.
13. Fei Zhang and Hao Yan, "DNA Self-Assembly Scaled Up," *Nature* 552 (2017): 34–35.
14. Grigory Tikhomirov, Philip Petersen, and Lulu Qiang, "Fractal Assembly of Micrometre-Scale Origami Arrays with Arbitrary Patterns," *Nature* 552 (2017): 67–71.
15. Klaus F. Wagenbauer, Christian Sigi, and Hendrik Dietz, "Gigadalton-Scale Shape-Programmable DNA Assemblies," *Nature* 552 (2017): 78–83.
16. Qiao Jiang et al., "Rationally Designed DNA-Origami Nanomaterials for Drug Delivery In Vivo," *Advanced Materials* 30, no. 40 (2018): 1804785.
17. Philip N. Dannhauser et al., "Durable Protein Lattices of Clathrin That Can Be Functionalized with Nanoparticles and Active Biomolecules," *Nature Nanotechnology* 10 (2015): 954–57.
18. It was Christian Anfinsen who proposed this theory in 1973. In a classic experiment, he showed that ribonuclease A could be completely unfolded by placing it in a solution of urea. Put back into a more biological environment, ribonuclease A protein spontaneously refolded and recovered its function. His proposal is known as *Anfinsen's dogma*. Christian B. Anfinsen, "Principles that Govern the Folding of Protein Chains," *Science* 181 (1973): 223–30.
19. A strong force driving the folding is the *hydrophobic effect*. Some parts of the amino acid chain are hydrophobic (they do not like to make bonds with water) so they come together to avoid water. In this way the protein folds and loses entropy, but the water molecules gain conformational freedom to increase their movement, providing an important drive (but not the only one) to deterministically find a final folded structure.

20. Robert Service, "This Protein Designer Aims to Revolutionize Medicines and Materials," *Science*, July 21, 2016: aaf5862.
21. In October 2000, the Folding@home progam was launched by Vijay Pande, a structural and computational biologist at Stanford University. Folding@home has found a very interesting way to increase computer power to search for conformations of proteins: use the processing resources of thousands of personal computers owned by volunteers (who have installed the software on their systems) when they are not using them. The project has pioneered the use of GPUs, PlayStation 3s, Message Passing Interface (used for computing on multi-core processors), and some Sony Xperia smartphones for distributed computing and scientific research. Volunteers can track their contributions on the Folding@home website, which makes volunteers' participation meaningful, even competitive, and often secures their long-term involvement. Folding@home is the best tool for sampling the way proteins might fold, but not for predicting the 3-D structure.
22. The key to the success of Rosetta was implementing an ingenious finding made in the 1990s by Chris Sander. Those were the early days of genome sequencing, and Sander and his collaborators thought that perhaps the DNA sequence in the gene that encodes for a protein could be useful for identifying two amino acids which, although they may be far away from each other in the protein sequence, are located close together in every folded protein structure. Sander hypothesized that the closeness of the two amino acids is important because they determine the folding structure. If this was true, these two amino acids would likely continue to be close to each other in the evolutionary history (or future) of a protein. If one of them mutated, the protein would not fold properly and would not work, but if they both mutated at the same time, perhaps the function could also evolve. So, if one had enough genomes, one could, in principle, search for those pairs of amino acids. If many such pairs could be found in a gene, they could be put into the computer program that calculates protein structures, to try to evaluate the possible structures that could work with those contact points. It is like finding the position of the "staples" that hold the protein together; the staples are put in place in the computer model, which then explores how the protein folds with those constraints. At the turn of the century, DNA genomic data started to pour into the databases, and Sander, working with Debora Marks at Harvard Medical School in Boston, developed a new statistical algorithm that succeeded in finding the coevolving pairs. They proved that with their technique they could constrain the position of a wide

variety of proteins for which there is no template for comparison. This achievement transformed the protein folding game, because combining the newly found "structural staple" positions with the methods developed by Baker and others around the world would reduce the amount of computational power and time required to resolve a structure.

23. Po-Ssu Huang, Scott E. Boyken, and David Baker, "The Coming of Age of De Novo Protein Design," *Nature* 537 (2016): 320–27.
24. Brian Kuhlman et al., "Design of a Novel Globular Protein Fold with Atomic-Level Accuracy," *Science* 302 (2003): 1364–68.
25. Michael Eisenstein, "Living Factories of the Future," *Nature* 531 (2016): 401–3.
26. Gabriel L. Butterfield et al., "Evolution of a Designed Protein Assembly Encapsulating Its Own RNA Genome," *Nature* 552 (2017): 415–20.
27. It started with Leonardo da Vinci in the sixteenth century and Galileo Galilei in the seventeenth: both studied the relationships between the anatomy (structure) and function of living organisms.
28. Jianwei Song et al., "Processing Bulk Natural Wood into a High-Performance Structural Material," *Nature* 554 (2018): 224–28.
29. Heechul Park et al., "Enhanced Energy Transport in Genetically Engineered Excitonic Networks," *Nature Materials* 15 (2016): 211–16.
30. In the future, when protein design becomes easier, such protein templates can be designed and produced to create any desired pattern. It might not be necessary to use viruses as a template.
31. Elisabet Romero et. al., "Quantum Coherence in Photosynthesis for Efficient Solar-Energy Conversion," *Nature Physics* 10 (2014): 676–82.

CHAPTER 3. NANO IN MEDICINE

1. Neanderthals might have known that, too, as has been recently discovered: Laura S. Weyrich et al., "Neanderthal Behaviour, Diet, and Disease Inferred from Ancient DNA in Dental Calculus," *Nature* 544 (2017): 357–61.
2. Youyou Tu received the Nobel Prize in Physiology or Medicine in 2015 for the discovery and the development of artemisinin, one of the most widely used medicines in the treatment of malaria. Tu got the inspiration for finding effective chemical compounds to treat malaria from ancient Chinese texts; the one that gave her the key was Ge Hong's *A Handbook of Prescriptions for Emergencies* (1574 CE). The description of the preparation—"A handful of qinghao immersed in 2 liters of water, wring out the juice and drink it all"—gave Tu the idea that avoiding heating was perhaps necessary

to extract the active compound. The fascinating story of her findings and the extraordinary circumstances of her research life are summarized in an article she wrote in 2011: Youyou Tu, "The Discovery of Artemisinin (Qinghaosu) and Gifts from Chinese Medicine," *Nature Medicine* 17 (2011): 1217–20.

3. Quoted from Pasteur's Sorbonne lecture of April 7, 1864. The full lecture is available online at http://www.rc.usf.edu/~levineat/pasteur.pdf.
4. Hazel de Berg, transcript of taped interview with Lord Howard Florey, April 5, 1967, National Library of Australia, Canberra, 9. From http://www.asap.unimelb.edu.au/bsparcs/exhib/nobel/florey.htm.
5. Ernst B. Chain, "The Chemical Structure of the Penicillins," 1945 Nobel Lecture, March 20, 1946, https://www.nobelprize.org/uploads/2018/06/chain-lecture.pdf.
6. Hodgkin remains the only British woman to have ever won a Nobel Prize in science, awarded in 1964 for chemistry. She also managed to resolve the structure of insulin after thirty years of work, and she is considered the founder of protein crystallography. But the British press reported her gender rather than her science. The *Daily Mail* led with the infamous "Oxford Housewife Wins Nobel," while the Telegraph wrote: "British Woman Wins Nobel Prize—£18,750 Prize to Mother of Three." Hodgkin supervised many other women who went on to become successful scientists, such as Clara Shoemaker, Rita Cornforth, Barbara Low, Cecily Darwin Littleton, Jenny Pickworth Gluster, Eleanor Dodson, and Judith Howard. One of them, Margaret Roberts, did not have a successful career in science, but she went on to become Prime Minister Margaret Thatcher.
7. Sumerian clay tablets, the Ebers Papyrus—an ancient Egyptian medical text (ca. 1550 BCE)—and Hippocrates in Greece (ca. 460–377 BCE) all mentioned the uses of willow to alleviate pain and fever.
8. Eichengrün was a Jew, and his participation was erased from the official records during Nazi Germany.
9. Robert D. Turner et al., "Cell Wall Elongation Mode in Gram-Negative Bacteria is Determined by Peptidoglycan Architecture," *Nature Communications* 4 (2013): 1496.
10. "Biodegradable Nanostructures with Selective Lysis of Microbial Membranes," *Nature Chemistry* 3 (2011): 409–14.
11. Larry Greenemeier, "Bursting MRSA's Bubble: Using Nanotech to Fight Drug-Resistant Bacteria," *Scientific American*, April 4, 2011.

12. To document the penetration of nanotechnology in the marketplace, the Woodrow Wilson International Center for Scholars and the Project on Emerging Nanotechnology launched the Nanotechnology Consumer Product Inventory (CPI) in 2005: www.nanotechproject.org/cpi/. Silver nanoparticles are the most widely used.
13. Alexander P. Richter et.al., "An Environmentally Benign Antimicrobial Nanoparticle Based on a Silver-Infused Lignin Core," *Nature Nanotechnology* 10 (2015): 817–23.
14. Xiaomei Dai et al., "Functional Silver Nanoparticle as a Benign Antimicrobial Agent that Eradicates Antibiotic-Resistant Bacteria and Promotes Wound Healing," *ACS Applied Materials and Interfaces* 8 (2016): 25798–807.
15. Sarel J. Fleishman et al., "Computational Design of Proteins Targeting the Conserved Stem Region of Influenza Hemagglutinin," *Science* 332 (2011), 816–21.
16. Merika Treants Koday et al., "A Computationally Designed Hemagglutinin Stem-Binding Protein Provides In Vivo Protection from Influenza Independent of a Host Immune Response," *PLoS Pathogens* 12 (2016): e1005409.
17. Bruno E. Correia et al., "Proof of Principle for Epitope-Focused Vaccine Design," *Nature* 507 (2014): 201–6.
18. Zev J. Gartner, Matthew W. Kanan, and David R. Liu, "Multistep Small-Molecule Synthesis Programmed by DNA Templates," *Journal of the American Chemical Society* 124 (2002): 10304–6.
19. Wenjing Meng et al., "An Autonomous Molecular Assembler for Programmable Chemical Synthesis," *Nature Chemistry* 8 (2016): 542–548.
20. Yasuhiro Matsumura and Hiroshi Maeda, "A New Concept for Macromolecular Therapeutics in Cancer Chemotherapy: Mechanism of Tumoritropic Accumulation of Proteins and the Antitumor Agent Smancs," *Cancer Research* 46 (1986): 6387–92.
21. Elvin Blanco, Haifa Shen, and Mauro Ferrari, "Principles of Nanoparticle Design for Overcoming Biological Barriers to Drug Delivery," *Nature Biotechnology* 33 (2015): 941–51.
22. Sarah DeWeerdt, "Bacteriology: A Caring Culture," *Nature* 504 (2013): S4–S5.
23. Coley's work is currently being revisited, and the company MBVax Bioscience is using modern laboratory techniques to produce "Coley fluid" for cancer treatment.
24. Karolina Palucka, "Q&A: Evidence Presenter," *Nature* 504 (2013): S9.

25. Antonio Lanzavecchia and Federica Sallusto, "Ralph M. Steinman 1943–2011," *Cell* 147 (2015): 1216–17.
26. Adjuvants are agents that enhance the efficacy of a vaccine.
27. Christine Gorman, "Cancer Immunotherapy: The Cutting Edge Gets Sharper," *Scientific American*, October 1, 2015.
28. Michael S. Goldberg, "Immunoengineering: How Nanotechnology Can Enhance Cancer Immunotherapy," *Cell* 161 (2015): 201–4.
29. Darrell J. Irvine, Melody A. Swartz, and Gregory L. Szeto, "Engineering Synthetic Vaccines Using Cues from Natural Immunity," *Nature Materials* 12 (2013): 978–90.
30. *Humoral immunity* involves antibodies, proteins, and peptides (small proteins) that are present in body fluids, or humors.
31. Matthias T. Stephan et al., "Therapeutic Cell Engineering with Surface-Conjugated Synthetic Nanoparticles," *Nature Medicine* 16 (2010): 1035–41.
32. Yiran Zheng et al., "*In Vivo* Targeting of Adoptively Transferred T-cells with Antibody- and Cytokine-Conjugated Liposomes," *Journal of Controlled Release* 172 (2013): 426–35.
33. Robert F. Service, "Nanoparticles Awaken Immune Cells to Fight Cancer," *Science News,* January 5, 2017: aal0581.
34. Hao Yin et al., "Non-Viral Vectors for Gene-Based Therapy," *Nature Reviews Genetics* 15 (2014): 541–55.
35. Jordan J. Green, Robert Langer, and Daniel G. Anderson, "A Combinatorial Polymer Library Approach Yields Insight into Nonviral Gene Delivery," *Accounts of Chemical Research* 41 (2008): 749–59.
36. Nicole Ali McNeer et al., "Nanoparticles That Deliver Triplex-Forming Peptide Nucleic Acid Molecules Correct F508del CFTR in Airway Epithelium," *Nature Communications* 6 (2015): 6952.
37. *Angiogenesis* is the process by which a tumor grows blood vessels to supply itself with nutrients in order to develop and survive.
38. Robert Langer and Judah Folkman, "Polymers for the Sustained Release of Proteins and Other Macromolecules," *Nature* 263 (1976): 797–800.
39. Robert Langer, "Biomaterials and Biotechnology: From the Discovery of the First Angiogenesis Inhibitors to the Development of Controlled Drug Delivery Systems and the Foundation of Tissue Engineering," *Journal of Biomedical Materials Research* 101A (2013): 2449–55.
40. Samir Mitragotri, Paul A. Burke, and Robert Langer, "Overcoming the Challenges in Administering Biopharmaceuticals: Formulation and Delivery Strategies," *Nature Reviews Drug Discovery* 13 (2014): 655–72.

41. Amy C. Richards Grayson et al., "Multi-Pulse Drug Delivery from a Resorbable Polymeric Microchip Device," *Nature Materials* 2 (2003): 767–72; Robert Farra et al., "First-in-Human Testing of a Wirelessly Controlled Drug Delivery Microchip," *Science Translational Medicine* 4 (2012): 122ra21.
42. Yue Lu et al., "Bioresponsive Materials," *Nature Reviews Materials* 2 (2016): 16075.
43. J. Yu et al., "Microneedle-Array Patches Loaded with Hypoxia-Sensitive Vesicles Provide Fast Glucose-Responsive Insulin Delivery," *Proceedings of the National Academy of Sciences of the USA* 112 (2015): 8260–65.
44. Hyunjae Lee at al., "A Graphene-Based Electrochemical Device with Thermoresponsive Microneedles for Diabetes Monitoring and Therapy," *Nature Nanotechnology* 11 (2016): 566–72.
45. Elie Dolgin, "Cancer Vaccines: Material Breach," *Nature* 504 (2013): S16–S17.
46. Ankur Singh et al., "An Injectable Synthetic Immune-Priming Center Mediates Efficient T-Cell Class Switching and T-Helper 1 Response against B Cell Lymphoma," *Journal of Controlled Release* 155 (2011): 184–92.
47. Sidi A. Bencherif et al., "Injectable Cryogel-Based Whole-Cell Cancer Vaccines," *Nature Communications* 6 (2015): 7556.

CHAPTER 4. RECREATING TISSUES AND ORGANS

1. In Greek and Roman mythology, Hydra, the serpentine water monster, would regrow one or multiple heads for every head that an enemy had chopped off. Prometheus, a favorite hero of this book, could also regrow his liver every night, after it had been eaten by an eagle sent by Zeus during the day. Modern versions abound, from the British series *Doctor Who* to uncountable comics and manga. Regenerative powers featured in video games, comics, TV series, and literature are compiled and classified on modern fan blogs such as "Regenerative Healing Factor" at SuperpowerWiki.
2. My favorite is the 1970s manga *Black Jack* by legendary author Osamu Tezuka, who was himself a medical doctor.
3. Jane Maienschein, "Controlling Life: From Jacques Loeb to Regenerative Medicine," *Journal of the History of Biology* 42 (2009): 215–30.
4. Philip Pauly, *Controlling Life: Jacques Loeb and the Engineering Ideal* (New York: Oxford University Press, 1987).
5. Pauly, 51.

6. Jacques Loeb, "Activation of the Unfertilized Egg by Ultra-Violet Rays," *Science* 40 (1914): 680–81.
7. Ross G. Harrison, "Observations on the Living Developing Nerve Fiber," *Proceedings of the Society for Experimental Biology and Medicine* 4 (1908): 140–43.
8. Leroy C. Stevens Jr and C. C. Little, "Spontaneous Testicular Teratomas in an Inbred Strain of Mice," *Proceedings of the National Academy of Sciences of the USA* 40 (1954): 1080–87.
9. Davor Solter, "From Teratocarcinomas to Embryonic Stem Cells and Beyond: A History of Embryonic Stem Cell Research," *Nature Reviews Genetics* 7 (2006): 319–27.
10. Gail R. Martin and Martin J. Evans, "The Morphology and Growth of a Pluripotent Teratocarcinoma Cell Line and Its Derivatives in Tissue Culture," *Cell* 2 (1974): 163–72.
11. James A. Thomson et al., "Embryonic Stem Cell Lines Derived from Human Blastocysts," *Science* 282 (1998):1145–47.
12. Charles A. Vacanti, "The History of Tissue Engineering," *Journal of Cellular and Molecular Medicine* 10 (2006): 569–76.
13. These pioneering studies were led by John Burke and Ioannis Yannas of the Massachusetts General Hospital and MIT.
14. Howard Green applied sheets of keratinocytes onto burn patients, and Eugene Bell seeded collagen gels with fibroblasts.
15. Robert Langer and Joseph P. Vacanti, "Tissue Engineering," *Science* 260 (1993): 920–26
16. Yilin Cao et al., "Transplantation of Chondrocytes Utilizing a Polymer-Cell Construct to Produce Tissue-Engineered Cartilage in the Shape of a Human Ear," *Plastic and Reconstructive Surgery* 100 (1997): 297–302.
17. In the late 1990s this picture went viral, circulated mainly via email, often without any text to explain it. Many people thought it was a fake. The picture provoked a wave of protests against genetic engineering, but in this experiment no genetic manipulation was performed. The ear was implanted. Even the strain of the mouse that displayed the ear is not genetically modified; it is the result of a spontaneous natural mutation.
18. "Tissue Engineering," Technologies, *Nature Biotechnology* 18 (2000): IT56—IT58.
19. Tal Dvir et al., "Nanotechnological Strategies for Engineering Complex Tissues," *Nature Nanotechnology* 6 (2011): 13–22.

20. Arvind Raman et al., "Mapping Nanomechanical Properties of Live Cells Using Multi-Harmonic Atomic Force Microscopy," *Nature Nanotechnology* 6 (2011): 809–14.
21. Jason W. Nichol and Ali Khademhosseini, "Modular Tissue Engineering: Engineering Biological Tissues from the Bottom Up," *Soft Matter* 5 (2009): 1312–19.
22. http://www.oxsybio.com/technology/.
23. Ashkan Shafiee and Anthony Atala, "Tissue Engineering: Toward a New Era of Medicine," *Annual Reviews of Medicine* 68 (2017): 29–40.
24. Kaitlyn Sadtler et al., "Design, Clinical Translation and Immunological Response of Biomaterials in Regenerative Medicine," *Nature Reviews Materials* 1 (2016): 16040.
25. Šárka Kubinová, "New Trends in Spinal Cord Tissue Engineering," *Future Neurology* 10 (2015): 129–45.
26. Jared T. Wilcox, David Cadotte, and Michael G. Fehlings, "Spinal Cord Clinical Trials and the Role for Bioengineering," *Neuroscience Letters* 519 (2012): 93–102.
27. Jia Liu et al., "Syringe-Injectable Electronics," *Nature Nanotechnology* 10 (2015): 629–36.
28. Dara Mohammadi, "The Lab-Grown Penis: Approaching a Medical Milestone," *Guardian*, October 4, 2014.
29. Ko-Liang Chen et al., "Bioengineered Corporal Tissue for Structural and Functional Restoration of the Penis," *Proceedings of the National Academy of Sciences of the USA* 107 (2010): 3346–50.
30. Okano is at the Institute of Advanced Biomedical Engineering and Science at Tokyo Women's Medical University.
31. Vivien Marx, "Tissue Engineering: Organs from the Lab," *Nature* 522 (2015): 373–77; "Cell Sheet–Based Myocardial Tissue Engineering: New Hope for Damaged Heart Rescue," Tatsuya Shimizu et al., *Current Pharmaceutical Design* 15 (2009): 2807–14.
32. Robert J. Morrison et al., "Mitigation of Tracheobronchomalacia with 3D-Printed Personalized Medical Devices in Pediatric Patients," *Science Translational Medicine* 7 (2015): 285ra64.
33. This result was reported by the group of Glenn Green at the University of Michigan (preceding reference), which 3-D–printed the splints using polycaprolactone (PCL), a biocompatible polyester that remains in place in vivo for 2–3 years before resorption by the body.

34. Hyun-Wook Kang et al., "A 3D Bioprinting System to Produce Human-Scale Tissue Constructs with Structural Integrity," *Nature Biotechnology* 34 (2016): 312–19.
35. Nadav Noor et al., "3D Printing of Personalized Thick and Perfusable Cardiac Patches and Hearts," *Advanced Science* (2019): 1900344.
36. Basma Hashmi et al., "Developmentally-Inspired Shrink-Wrap Polymers for Mechanical Induction of Tissue Differentiation," *Advanced Materials* 26 (2014): 3253–57.
37. Sangeeta N. Bhatia and Donald E. Ingber, "Microfluidic Organs-on-Chips," *Nature Biotechnology* 32 (2014): 760–72.
38. Gordana Vunjak-Novakovic et al., "HeLiVa Platform: Integrated Heart-Liver-Vascular Systems for Drug Testing in Human Health and Disease," *Stem Cell Research and Therapy* 4 (2013): S8.
39. Jannick Theobald et al., "Liver-Kidney-on-Chip To Study Toxicity of Drug Metabolites," *ACS Biomatererials Science and Engineering* 4 (2018): 78–89.
40. David B. Kolesky et al., "Three-Dimensional Bioprinting of Thick Vascularized Tissues," *Proceedings of the National Academy of Sciences of the USA* 113 (2016): 3179–84.
41. Johan U. Lind et al., "Instrumented Cardiac Microphysiological Devices via Multimaterial Three-Dimensional Printing," *Nature Materials* 16 (2017): 303–8.
42. Aliya Fatehullah, Si Hui Tan, and Nick Barker, "Organoids as an *In Vitro* Model of Human Development and Disease," *Nature Cell Biology* 18 (2016): 246–54.
43. Marx, "Tissue Engineering: Organs from the Lab."
44. Sung-Jin Park et al., "Phototactic Guidance of a Tissue-Engineered Soft-Robotic Ray," *Science* 353 (2016): 158–62.
45. The optical sensitivity is achieved using *optogenetics*. Optogenetics uses a viral vector to insert a gene into the host cell that then expresses a protein at the cell surface that is sensitive to light.
46. Elizabeth Pennisi, "Robotic Stingray Powered by Light-Activated Muscle Cells," *Science* News, July 7, 2016: aaf5835.
47. Michel Foucault, *The History of Sexuality*, vol. 1: *The Will to Knowledge* (London: Penguin, 1998), 140.

CHAPTER 5. CONCLUSIONS

1. This has been extensively studied by K. Eric Drexler in his book *Radical Abundance: How a Revolution in Nanotechnology Will Change Civilization* (New York: PublicAffairs, 2013).
2. Anna Nowogrodzki, "Inequality in Medicine," *Nature* 550 (2017): S18–S19.

EPILOGUE. BIOLOGY BECOMES PHYSICS: OUR COMING OF AGE AS A TECHNOLOGICAL SPECIES?

1. http://www.matterforall.org/
2. Hillary Sutcliffe, "5 Lessons from the Past for the Fourth Industrial Revolution," World Economic Forum, February 23, 2017, https://www.weforum.org/agenda/2017/02/lessons-from-nanotech-for-the-4th-industrial-revolution/.
3. K. Eric Drexler, *Radical Abundance: How a Revolution in Nanotechnology Will Change Civilization* (New York: PublicAffairs, 2013).
4. An archetypal example is Doraemon. The cat-robot from the future has been part of children's lives in Japan since the first manga series was published in 1969. The manga were followed by a television series that is now watched by children all over the world. Doraemon arrives from the twenty-second century to help Nobita Nobi with his problems with friends, family, and school. Like its predecessor, Astro Boy, the robot is a force for good, helping humans, making their life more fun, more comfortable and interesting.
5. Anna Dickie, "Toshiyuki Inoko in Conversation," *Ocula* magazine, January 13, 2014.
6. Ashlee Vance, "Japan's Obsessive Robot Inventors Are Creating the Future, Episode 8: The Tech Industry in Japan Has Awakened through Androids, Insane Art and Quirky Inventors," *Bloomberg Businessweek*, October 26, 2016.
7. The words of Francesco Petrarca (Petrarch), founder of humanism and initiator of the Renaissance.

Index